Introduction to Modeling Cognitive Processes

Introduction to Modeling Cognitive Processes

Tom Verguts

The MIT Press
Cambridge, Massachusetts
London, England

© 2022 The Massachusetts Institute of Technology

All rights reserved. No part of this book may be reproduced in any form by any electronic or mechanical means (including photocopying, recording, or information storage and retrieval) without permission in writing from the publisher.

The MIT Press would like to thank the anonymous peer reviewers who provided comments on drafts of this book. The generous work of academic experts is essential for establishing the authority and quality of our publications. We acknowledge with gratitude the contributions of these otherwise uncredited readers.

This book was set in Stone Serif and Stone Sans by Westchester Publishing Services. Printed and bound in the United States of America.

Library of Congress Cataloging-in-Publication Data

Names: Verguts, Tom, author.
Title: Introduction to modeling cognitive processes / Tom Verguts.
Description: Cambridge, Massachusetts : The MIT Press, 2022. | Includes bibliographical references and index.
Identifiers: LCCN 2021016752 | ISBN 9780262045360 (hardcover)
Subjects: LCSH: Cognition—Data processing. | Cognition—Computer simulation. | Cognitive learning. | Cognitive neuroscience. | Human information processing.
Classification: LCC BF311 .V4467 2022 | DDC 153.9/3—dc23
LC record available at https://lccn.loc.gov/2021016752

10 9 8 7 6 5 4 3 2 1

Contents

Preface and Acknowledgments   ix

1 What Is Cognitive Modeling?   1

The Use of Models   1
Time Scales of Modeling   5
Striving for a Goal   6
Optimization   8
TensorFlow   13
Minimizing Energy or Getting Groceries   14

2 Decision Making   17

Minimization in Activation Space   17
A Minimal Energy Model   21
Cooperative and Competitive Interactions in Visual Word Recognition   25
The Hopfield Model   27
Harmony Theory   30
Solving Puzzles with the Hopfield Model   31
Human Memory and the Hopfield Model   32
The Diffusion Model   33
The Diffusion Model in Psychology   35

3 Hebbian Learning   37

The Hebbian Learning Rule   37
Biology of the Hebbian Learning Rule   40
Hebbian Learning in Matrix Notation   41
Memory Storage in the Hopfield Model   44
Hebbian Learning in Models of Human Memory   48

4 The Delta Rule   53

The Delta Rule in Two-Layer Networks   53
The Geometry of the Delta Rule   58

The Delta Rule in Cognitive Science   61
The Rise, Fall, and Return of the Delta Rule   66

5   Multilayer Networks   69

Geometric Intuition of the Multilayer Model   69
Generalizing the Delta Rule: Backpropagation   72
Some Drawbacks of Backpropagation   74
Varieties of Backpropagation   76
Networks and Statistical Models   82
Multilayer Networks in Cognitive Science: The Case of Semantic Cognition   83
Criticisms of Neural Networks   85

6   Estimating Parameters in Computational Models   89

Parameter Space Exploration   89
Parameter Estimation by Error Minimization   91
Parameter Estimation by the Maximum Likelihood Method   92
Applications   99

7   Testing and Comparing Computational Models   107

Model Testing   108
Model Testing across Modalities   114
Model Comparison   116
Applications of Model Comparison   120

8   Reinforcement Learning: The Gradient Ascent Approach   123

Gradient Ascent Reinforcement Learning in a Two-Layer Model   124
An N-Armed Bandit   126
A General Algorithm   127
Backpropagating RL Errors   129
Three- and Four-Term RL Algorithms: Attention for Learning   130

9   Reinforcement Learning: The Markov Decision Process Approach   133

The MDP Formalism   134
Finding an Optimal Policy   138
Value Estimation   138
Policy Updating   143
Policy Iteration   143
Exploration and Exploitation in Reinforcement Learning   143
Applications   145
Combining Gradient-Ascent and MDP Approaches   149
Reinforcement Learning for Human Cognition?   151
Open AI Gym   152

Contents

10 Unsupervised Learning 153

Unsupervised Hebbian Learning 153
Competitive Learning 156
Kohonen Learning 158
Auto-Encoders 161
Boltzmann Machines 162
Restricted Boltzmann Machines 166

11 Bayesian Models 173

Bayesian Statistics 173
The Rational Approach 179
Bayesian Models of Cognition 182

12 Interacting Organisms 191

Social Decision Making 192
Combining Information 193
Game Theory 193
Cultural Transmission and the Evolution of Languages 198
To Conclude 201

Conventions and Notation 203
Glossary 205
Hints and Solutions to Select Exercises 207
Notes 217
References 219
Index 243

Preface and Acknowledgments

Cognitive science and cognitive neuroscience have witnessed an explosion of novel theories, tools, and data in recent decades. This increasing sophistication requires conceptual tools to make sense of it all. Computational modeling is just such a tool; it allows turning verbal theories into precise formulations, thus aiding both the integration of behavioral, electroencephalography (EEG), functional magnetic resonance imaging (fMRI), and other kinds of data and testing of the theory. As a result, computational modeling is omnipresent in modern cognitive science and cognitive neuroscience. At the same time, the general public is increasingly interested in the relation between cognition and computation; witness the popularity of topics like deep learning.

Oddly, however, there are very few textbooks targeted at an audience of cognitive neuroscientists. Note: I will use the term "cognitive neuroscience" in a very inclusive manner, encompassing cognitive neuroscience, cognitive science, neuroscience, psychology, and all its intersections. Most books on the market are targeted toward computer scientists or engineers. One excellent book (McLeod, Plunkett, & Rolls, 1998) is targeted toward a cognitive neuroscience audience; I have used this for several years in my modeling cognitive processes class. However, that book is now becoming outdated. There are good more recent books targeted to the same cognitive neuroscience audience (Farrell & Lewandowsky, 2018), but they tend to emphasize statistical model analysis, whereas I wanted a stronger focus on model building in my class. This prompted me to write this textbook. It also covers model analysis (in chapters 6 and 7), but the emphasis is on model building (in all the other chapters). The audience is meant to be inclusive: it caters to novice readers, but also to active researchers who want to refresh their memory or study a field of modeling that they have less experience with.

Readers of the research literature typically find a chaotic and overwhelming variety of models on the market. This can make it hard to see the proverbial forest for the trees. With this text, I had the specific goal to provide the reader a unified perspective on modeling. To keep it concise and self-contained, I had to exclude many interesting models and modeling approaches. Relatedly, I do not only cover the latest hits (although some of these are discussed as well), but intend to show that the same modeling principles have been recycled throughout the last decades. I hope the text is a useful starting point for readers to dig into the more specialized literature themselves.

Thematically, after chapter 1, the book starts with models of decision making (chapter 2), followed by supervised learning (chapters 3–5), statistical model analysis (chapters 6 and 7), reinforcement learning (RL; sometimes called *semisupervised learning*) (chapters 8 and 9), unsupervised learning (chapter 10), and finally Bayesian (chapter 11) and social-interaction (chapter 12) models. I use a subset (chapters 1–7, 9, and 10) in my modeling cognitive processes (MCP) class at Ghent University. Note that the division of material across chapters is model-based rather than topic-based (such as a typical "introduction to psychology" book would be). For example, the topic of language appears in different chapters (e.g., chapters 4, 5, and 12). The only exception to this rule is chapter 12, which describes various models of social interactions. Although the book is intended as an integrated whole, it is possible to read the chapters individually. However, I do recommend reading chapter 1 first, and to read chapters 4–5 and chapters 6–7 together due to their thematic overlap.

The language of modeling is mathematical, but don't be afraid. Anyone with a high-school background and a willingness to invest some effort should be able to follow most of this book, and at least the main arguments of each chapter. Some of the material is slightly more advanced, but this material can be skipped without danger of losing the main argument. There are also theoretical exercises throughout the book to help you develop your modeling skills. Exercises vary a lot in difficulty, so that researchers with different backgrounds can profit. The more advanced exercises are indicated with an asterisk (*). Hints and solutions to the exercises of the book are provided at the end.

In addition, the book contains a glossary with some of the key technical terms that I use. The definitions provided here are intended to refresh your memory; they are typically not sufficient to understand a concept. You cannot understand the derivative of a function just by reading a two-line definition. If you need to learn about or refresh some basic concepts, there are excellent websites available to do just that. I recommend Xaktly.com, which offers very good, graphical, and intuitive tutorials on many mathematical concepts.

One of the advantages of a mathematical model is that ideally, it is precise enough to be implemented in a computer program. A lot of effort in modeling, therefore, is also directed toward computer code implementation. For the code in this book, I used the free Python language, with the Spyder editor in the Anaconda environment. Visit Anaconda.org (www.anaconda.org) for download and tutorial on this language. There are several good tutorials on Python available. For a thorough theoretical introduction, I recommend Punch and Enbody (2014) or my own website (https://sites.google.com/view/pp02psychopy), which focuses on programming experiments but also introduces Python in a practical manner.

In this book, I refer to some packages that you can download and install. It is advised to create a virtual environment in Anaconda for each collection of packages that you use together. In this way, you can install in that environment the optimal versions of basic packages such as numpy or scipy without interference from other versions. For

Preface and Acknowledgments

example, in my Anaconda installation, I have an environment for using AI Gym and TensorFlow, and another environment for using PsychoPy (Peirce, 2007).

Modeling can only be learned by doing, so exercise and exploration are indispensable. For this reason, the code for generating all figures and the coding exercises are also provided (as Python .py files) on GitHub. GitHub is a platform to share computer code. Here, each coding project (such as a chapter in this book) is conceptualized as an online tree consisting of the different programs belonging to the project. The owners of a tree can directly change code. If other users have suggestions for code changes, they can create a branch in the tree, which the owners can then inspect and subsequently either reject ("cut off the branch"), or merge with the tree. In this way, joint code creation for a project is possible. To get started, you can just download the code for your inspiration and exploration. But if you have suggestions, do feel free to add a branch! The Github repository for this book can be found at

https://github.com/CogComNeuroSci/modeling-master/tree/master/code%20by%20 chapter.

If you want to use this book for teaching, I can also send you extra exercises, tests, and their solutions that I use in class. Some extra material (currently only color pictures) is also posted at https://sites.google.com/view/mcp-website; in the book, this is referred to as the "MCP website."

I thank Mehdi Senoussi and Pieter Huycke for providing excellent help in teaching the course that started this textbook, for feedback on the different chapters, and for developing several exercises to accompany the text. I thank the MCP students for suffering through the first years where I tried out a preliminary draft of this book, as well as for providing useful comments on several passages. Thanks are also due to Esther De Loof for Python inspiration, and to Anna Marzecova and Jonas Simoens for teaching support. For comments on parts of an earlier draft, I thank Kobe Desender, Wim Fias, Rob Hartsuiker, Clay Holroyd, and Francis Tuerlinckx. I thank Martin Butz and three anonymous reviewers for their feedback; and Philip Laughlin and his team from the MIT Press for supporting the project and providing practical advice. All remaining errors in the text are mine. Finally, I thank my partner, Ann, and my daughters, Laura-Line and Yente, for being there for me.

On the day that I write these words (in April 2021), the vaccination program against the SARS-CoV-2 virus that led to a worldwide pandemic is in full swing. It has been a difficult year for many people and for many reasons, but also for those who had to learn a new skill, such as modeling. This fact can be understood from a modeling perspective: As will become clear from reading the book, the most efficient way to train a model is to let it do something first, followed by looking at feedback on its performance. Providing feedback was more difficult due to the necessity to keep social distance, thus impairing the learning experience. This is one lesson to take away from the crisis: feedback is crucial for learning any skill, and writing a book is no exception. I therefore welcome any comments the reader of this book may have, regardless of whether you read it as a teacher, a student, a researcher, or just for fun.

1 What Is Cognitive Modeling?

Don't just stand there, optimize something!
—Daniel S. Levine

The Use of Models

What is cognitive modeling? For several years, I have attempted to answer this question at family and cocktail parties. It's not so easy to figure this out because the term has different meanings in different contexts, including ecology, astronomy, meteorology, and economics. So let's start by making clear how it will be used in this book. In cognitive neuroscience, modeling is a tool to help construct better theories of cognition and behavior. We have empirical tools in psychology, such as response time measurement, electroencephalogram (EEG) measurement, and questionnaires. However, modeling is a conceptual rather than an empirical tool. In particular, it provides a formal language that helps build more precise theories.

Why do we need this conceptual tool? Models have various advantages. First, they allow for making novel predictions that don't obviously follow from the theory. A schoolbook example of modeling concerns Newton's theory of gravity (an example borrowed from John Kruschke[1]). At an intuitive (nonprecise) level, the theory of gravity states that due to a force called "gravity," each pair of objects in the universe is attracted to each other. In itself, this is not so useful or interesting, but we can make it interesting by formalizing it into a model. Thus, Newton's model also specifies that gravity is proportional to the product of the masses of the two objects, and further, that gravity is inversely proportional to the distance between the objects. From these two principles (and some other laws that I'll ignore for now), one can predict that all objects will fall toward the Earth at the same acceleration (and thus speed). Would anyone be able to make this prediction without the formal model?

Second, models allow integration of existing data in a well-organized conceptual framework. The theory of gravity allows formalizing the finding (first discovered by

Galilei) that the movement of a pendulum always has the same frequency (movement speed), which allowed the construction of precise clocks. The theory also helps us understand how tides work; why the Moon rotates around the Earth; the Earth around the Sun; the Sun in our galaxy; and many more phenomena. Quite a heavy lift for such a simple concept! Our understanding of the world would simply not be the same without the theory of gravity—and the theory of gravity would not be the same without the conceptual tool of modeling.

Obviously, I am not going to talk about physics in this book. Yet, in cognitive neuroscience, modeling is also an extremely useful tool; for a philosophical and historical background to cognitive modeling, see Butz and Kutter (2017). You may have heard about the replication crisis in psychology, meaning that many empirical effects in psychology could not be replicated (Open science collaboration, 2015). Several measures have been taken to address it, including preregistration of study designs and Bayesian statistics. Although such measures are definitely relevant and worthwhile, they have tended to concentrate on improving the statistical methods. However, some have argued that the deeper crisis resides not only in the methods, but also in the lack of theory that psychologists develop (Muthukrishna & Henrichs, 2019; Oberauer & Lewandowsky, 2019; van Rooij & Baggio, 2020). Modeling is a tool to help developing such theories. I hope to demonstrate this in this book.

A crucial concept in cognitive modeling is levels of modeling. Modeling can be done at a "high" level, describing interactions between large units; or at a "low" level, where one considers interactions between smaller units. These levels are illustrated on the y-axis of figure 1.1. This is called the "spatial scale" in figure 1.1, in the sense that higher levels denote interactions between larger spatial units.

Starting at the top, one can consider interactions at the social level, between individuals (social agents), or even between entire groups of individuals. For example, modeling at this level demonstrates the precise conditions under which cooperation between individuals pays off more than defection (Nowak, 2006). A general conclusion from this modeling work is that there is always a strong attraction toward defection, but sufficiently strong interactions between individuals can make cooperation stronger, thus leading to profitable interactions (see chapter 12).

Another example of a social-level model is the Lotka-Volterra model, which considers interactions between predators (e.g., lions) and prey (e.g., deer). The model makes the rather intuitive assumption that local interactions between predators and prey increase the number of predators, but interactions decrease the number of prey. Formalizing this notion, it turns out that both population sizes will oscillate across time, alternating between periods with more and fewer animals. Furthermore, they oscillate in a synchronized way; if the number of prey increases, then the number of predators follows suit. Nothing or nobody needs to coordinate this synchrony—it follows from the local interactions among the animals. And we can understand this consequence only by turning the initial assumptions into a formal model.

Descending on the spatial scale, one can consider interactions within a single person or within a single brain. Here, one may be interested in modeling how cognitive modules interact to explain vision (Poggio & Bizzi, 2004), motor processing (Wolpert et al., 1995), or indeed any other cognitive function. As another example, Jonathan Cohen, Matthew Botvinick, and several colleagues have developed an influential series of models about how humans and other primates implement cognitive control, which is the ability to control one's own cognitive processes, such as to suppress one's instantaneous urges in favor of more appropriate responses (Botvinick et al., 2001). For example, a canonical task for measuring cognitive control is the Stroop task, in which subjects must read the ink color (say, green) of words that state a color (say, "orange"); people have the automatic tendency to read the word instead (Google "Stroop" and try the test online). It is thought that the extent to which people can suppress their urge to name the word instead of the color is a measure of cognitive control.

A typical interpretation of the cognitive processes involved in solving this task goes as follows: One cognitive module processes words, one cognitive module processes colors, and a third one processes verbal or manual actions. The anatomical connectivity structure between the word-processing module and the response module is much stronger than between color-processing and response modules because pronouncing words is a much more common, and thus more practiced, task than naming colors. Thus, the challenge in the Stroop task is that, despite its weaker anatomical connectivity, the color module must dominate the response module to solve the task correctly. For this purpose, a module is postulated that monitors whether the other three modules are acting appropriately. If they are not, the monitoring module intervenes to make sure that the color module has a larger impact on the response module. Several cognitive control models address this challenge and have this basic structure (Brown & Braver, 2005; Verguts & Notebaert, 2008). At the neural level, the monitoring module is typically thought to involve both medial and lateral frontal cortices, possibly in interaction with basal ganglia structures.

At a still lower level, further descending on the y-axis of figure 1.1, researchers study the interactions between single neurons (Wong & Wang, 2006). But there is no need to stop at the single-neuron level: modeling can (and indeed does) continue to any level where one has data.

Figure 1.1 is an approximation: Even only at the brain level, eleven spatial levels of investigation have been distinguished (Sejnowski, 2020). Confusion sometimes arises when people do not clearly distinguish among these levels. Thus, an often-heard criticism is that a specific model is not biologically plausible (but often lacking details about what specifically is implausible). However, whether biological plausibility is relevant and to what extent depend entirely on the question that the modeler is attempting to address. Whether a model is biologically plausible has different meaning and relevance for each of these models. For example, the social-interaction model mentioned here considers interactions between very simple agents. These conceptual agents do not consist of biological cells with dendrites, axons, glutamate receptors,

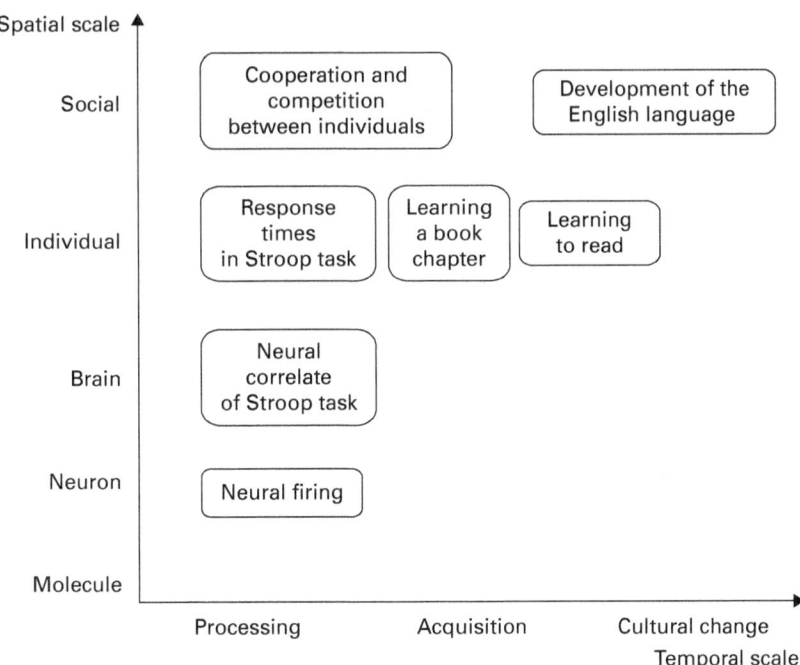

Figure 1.1
Levels (or scales) of modeling. The spatiotemporal scale of several real-world phenomena that can be modeled are indicated on the graph.

sodium-potassium pumps, and so on. In this sense, the model and the agents that it considers are not biologically plausible. However, this doesn't matter at this level of modeling. In fact, it would detract massively from the goal of modeling interactions between individuals to include all this biological detail. It is precisely *because* some biological details are ignored that it is at all possible to formulate useful models at the social level. In contrast, a neural model that aims to understand communication at the cell membrane requires the concept of a sodium-potassium pump if it is of any use for understanding such communications. For this reason, modeling is to a large extent an art: it requires being able to formulate a useful level of abstraction, with "useful" meaning that the model allows for integrating and understanding data at that specific level and allowing the derivation of novel empirical tests. But what exact level of detail is useful will differ across levels. Ideally, we should also be able to know how to connect explanations across different levels (e.g., from neuron to brain to social level). However, just as in other sciences, such cross-level comparisons are typically very hard.

As another example, several authors have suggested that mental diseases, such as schizophrenia, are due to altered connectivity in the brain (Bassett et al., 2008). For example, using network theory, van den Heuvel et al. (2013) found that the highly connected brain hubs in schizophrenic patients were less mutually connected than the same regions in control patients. In constructing and testing this model, many neural details were deliberately ignored. This is in some sense a loss. However, it did allow the modelers to concentrate on the connectivity structure between the neural hubs and how they are impaired in schizophrenia. Again, it is precisely *because* they ignored some details that they could achieve the right level of abstraction, focus on connectivity among neural areas, and derive useful empirical predictions from the abstract connectivity structure.

This book mostly concerns cognitive modeling; it primarily targets the middle part of the vertical axis of figure 1.1. It addresses cognition at the level of the individual subject or individual brain, and it typically considers interactions among several neural modules within a single subject (or brain). Note that also the term "module" can be understood at different levels; it can refer to a neural column, a brain region, or a cognitive module, depending on the level of granularity of the investigation.

This doesn't mean that the other levels are irrelevant. In fact, in cognitive modeling, one is typically interested in a few levels at the same time. The models I consider here aim to be cognitively plausible; that is, they must account for behavioral data such as response times and accuracies of subjects in behavioral experiments. But at the same time, the models should ideally also be neurally plausible; that is, they should be consistent with systems-level neuroscience or sometimes single-cell neuroscience. Different models will place different emphases on each of these levels. Obviously, in cognitive modeling, the cognitive plausibility question is a very important one; but it is not the only one.

Time Scales of Modeling

In addition to levels (spatial scales) of modeling, one can identify different time scales of modeling. In line with Christiansen and Chater (2016a), I distinguish at least three time scales. The fastest time scale considered in this book is that of cognitive processing and human behavior. At this time scale, people utter sentences, remember telephone numbers, play tennis, and so on. This is the time scale of the Stroop model previously discussed (see the horizontal axis of figure 1.1) and the time scale that will be considered in chapter 2. A slightly slower time scale is that of knowledge acquisition. This is the time scale at which new information (e.g., learning a book chapter) or skills (e.g., learning to read) are acquired. A lot of interest in the cognitive modeling literature has been devoted to this time scale. I accordingly direct most of my attention in this book (i.e., chapters 3–5 and 8–11) to this time scale. Finally, the slowest time scale considered here is that of cultural change. At this time, a human language transfers across generations of humans. Because this is the least-studied time scale (at least in cognitive neuroscience), only one chapter (chapter 12) considers it. And an even slower time scale is that of genetic change, which is not discussed in this book. An important theoretical goal is formulating common principles across temporal scales (e.g., processing and acquisition; Rao & Ballard, 1999); or cultural change and genetic change (Dawkins, 1976). To understand cognition and behavior, we eventually have to know how all those time scales relate to one another.

Striving for a Goal

The evolutionary biologist Theodosius Dobzhansky (figure 1.2) famously said, "Nothing in biology makes sense except in the light of evolution." If one accepts that cognition is a natural phenomenon, then this principle applies to cognition too. From this perspective, the function of cognition is to optimize an agent's interactions with the world. For example, in reasoning, arguments are not generated by humans applying a set of argumentation rules in an arbitrary fashion. Humans reason in order to convince others (Mercier & Sperber, 2011), which presumably optimizes the agent's interactions. More generally, it has been argued that the real reason for an organism to have a brain is so it can adaptively move within its environment (Wolpert, 2012). From this perspective, cognition is just a sophisticated way of choosing the right movement.

From a biological perspective, whether cognitive agents arrive at true conclusions is less relevant than whether the conclusions are useful for the agent. More generally, I postulate that humans act because they think it is useful for them in some way. In other words, they act because they are *motivated* to act. From an evolutionary perspective, one might say that organisms have evolved over millions of years to be maximally adaptive. Therefore, what they do at the current time can probably be considered as striving for adaptivity in some sense. This is the founding assumption of the field of

Figure 1.2
Theodosius Dobzhansky (1900–1975), one of the founders of the "modern synthesis" theory of biology (and inventor of great quotes). Reprinted with permission from the American Philosophical Society.

What Is Cognitive Modeling? 7

behavioral ecology (Shennan, 2002), and it can account for several of the biases that have been documented in the social psychology literature (Haselton et al., 2009). In biology, it has been debated to what extent the assumption is valid, but I'll bypass that discussion here and simply explore how far the assumption takes us in unifying models of cognitive processing.

One can formulate this intuition of adaptivity using the central concept of optimization. This is so central that I will attach a name to it: the *optimization principle*. The term "optimization" means that one maximizes or minimizes a function of some variable. For example, consider the function $y = f(x) = (x - 1)^2$, shown in figure 1.3. This function is very easy to minimize by choosing $x = 1$. Indeed, the function $y = f(x)$ reaches its minimal value of zero when $x = 1$; all other values would lead to values $f(x) > 0$. If one wanted to maximize this function instead, the solution would be at the boundary of parameter space: indeed, only $x = -\infty$ and $+\infty$ maximize this function.

Bringing these concepts together, I can formulate a core idea of this book: *human behavior is directed toward achieving some goal*. This can be formalized as attempting to optimize some function. The idea that human behavior is always directed toward achieving a certain goal can hardly be called a novel insight. It pervades motivational psychology (Deci & Ryan, 2008), and more broadly, literature, art, and indeed virtually any conversation about humans and their behavior. However, what may be surprising

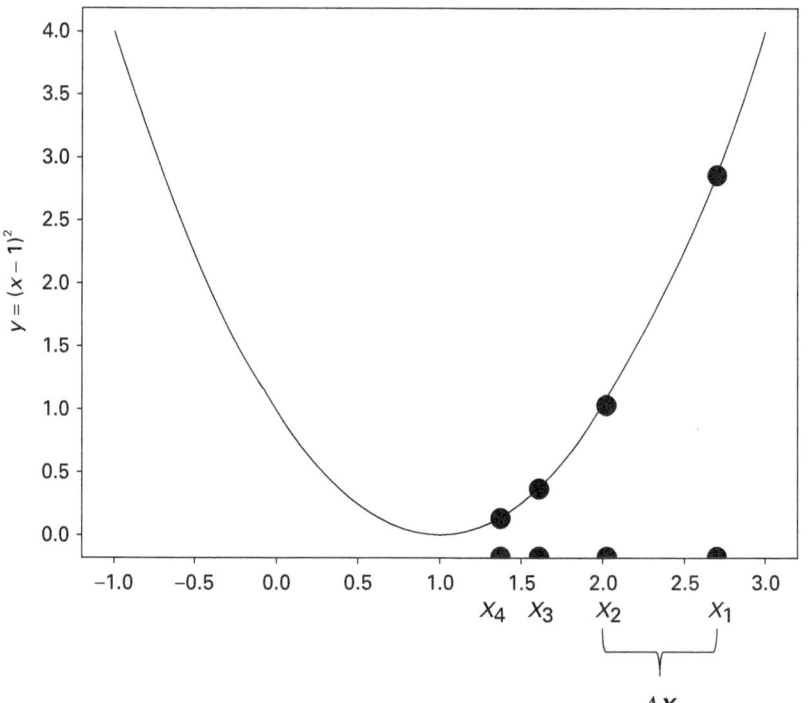

Figure 1.3
The steepest descent on the function $f(x) = (x - 1)^2$.

to you is this concept of motivation having such a central position in cognitive modeling. The person with just a casual experience of modeling often comes away from that experience thinking that cognitive models are mainly about associations between stimuli and responses, that cognitive models are a sophisticated form of regression, or that they are games, intended to satisfy the intellectual curiosity of the modeler. Models can be all of those things. But crucially, most cognitive models have a goal function at their core, which the model is motivated to optimize (Lieder & Griffiths, 2020). There are some exceptions to this rule; not all models can be considered as being involved in optimizing something. But for a first approximation, we can say that models attempt to reach a goal; and attempting to reach a goal can be formalized as optimizing a mathematical function.

Is the optimization principle really true? This is again hard to tell, but there are at least two reasons to accept it anyway. The first reason is exemplified by the quote of Dobzhansky mentioned earlier; given that humans are the descendants of animals across millions of years of evolution, it makes sense to assume that they attempt to be as adaptive as possible. The second, more pragmatic reason is that an extensive mathematical and statistical apparatus has been developed for optimizing functions. Application of the optimization principle allows us to exploit that apparatus and formulate models of behavior that are grounded in goal-directness. Also, note that the central claim is *not* that behavior is optimal, merely that humans attempt to optimize a goal. This is quite a difference: a funny but slightly misguided criticism of optimality theory is that a chess player gains nothing with the advice, "Simply optimize your probability of winning the game." This unhelpful statement assumes that the chess player knows how to do this, but the difficulty (and fun) of the game, of course, is that he or she does not know for sure and merely attempts to maximize this probability. Instead, the central claim is that much of (and perhaps all) cognition and behavior can be understood as attempts to climb or descend the function that is being optimized. And, to reiterate, a standard mathematical apparatus is available to formalize that notion. Let's now delve a bit more into that apparatus.

Optimization

The function that is optimized is called a *goal function*. The idea here is that the organism's goal is to optimize this function. The concept of having a goal function that is optimized is used in several fields. In economics, as developed by Adam Smith, Jeremy Bentham, and other founders of that discipline, consumers are considered to be *homo economicus*, whose sole interest is to maximize their own well-being. Well-being is typically (and conveniently) operationalized by the monetary value that a consumer can receive from each of a number of options. As a simple example, suppose that a farmer has the option to plant 1 apple tree at a cost of 10€, with an expected number of 50 apples, for which she receives 1€ each. Alternatively, she can plant 2 trees at a cost of 7€ each, with an expected number of 40 apples each, but in this case she gets only .70€ per apple (perhaps due to apple inflation). Simple arithmetic then demonstrates that the expected return in the first case is $50 \times 1 - 10 = 40€$; and in the second case, $2 \times 40 \times .70 - 2 \times 7 = 42€$. Optimality dictates, therefore, that she should plant two trees.

This principle also forms the basis of expected utility theory (von Neumann & Morgenstern, 1944; see also the discussion of game theory in chapter 12), which generalizes the notion of monetary value optimization. According to expected utility theory, each object x has a utility (or value) of $v(x)$, where $v(.)$ is a function measuring the value of object x. Subjects choose the object x with maximal $v(x)$. More generally, suppose that we have an option bundle (x_1, p_1, x_2, p_2), which means that object x_i can be obtained with probability p_i. The expected value of this option bundle then equals $p_1 v(x_1) + p_2 v(x_2)$. For example, should you play a lottery that costs 2€ but delivers 100€ with 1% probability? Suppose that the value of each monetary amount x is simply the amount itself: $v(x) = x$. Then the no-play option delivers you 0€; but the play option delivers you $0.99 \times (-2) + 0.01 \times 100 = -0.98$. Because the no-play option has a higher value $(0 > -0.98)$, you should not play in this particular situation with this set of assumptions. From this perspective, life is a search for the best option bundles available in the world.

Recent economic theories question the assumptions of the standard model, according to which humans exclusively attempt to maximize their own monetary value, and instead emphasize that humans value cooperation and social interactions more generally for achieving economic and psychological well-being (Nowak, 2006; Raworth, 2017). As a simple example, consider the ultimatum game. Here, player A gets to divide a sum of money between herself and player B. Player B knows how much money is at stake and can decide to either accept the offer of player A or reject it. In the latter case, neither player gets anything. From a purely financial perspective, player B should accept any offer larger than zero. For example, if player A gets to divide 100€ and offers 1€ to player B, there is no financial gain whatsoever to reject. And yet human players usually reject such very unfair offers (Mesoudi, 2011). Thus, the goal function that humans optimize is clearly not of a purely financial nature, but it does include a consideration of social fairness. Importantly, though, such findings and alternative theories do not question the idea that humans strive to optimize some utility (or goal) function. They just consider broader goals, incorporating social interactions or fairness, environmental considerations, and other factors in addition to monetary gain.

Perhaps more problematic for expected utility theory is that systematic deviations from its predictions have been observed in the last several decades. For example, in their classic paper, Kahneman and Tversky (1979) give the following example: Subjects can choose between option A = (4,000, 0.8) (meaning 4,000 with 80% probability) or option B = (3,000, 1) (meaning 3,000 for sure). In this case, subjects generally prefer option B over option A, implying that $v(3,000) > 0.8 \times v(4,000)$. However, if subjects can choose between option bundles C = (4,000, 0.2) and D = (3,000, 0.25), they go for bundle C, implying that $0.2 \times v(4,000) > 0.25 \times v(3,000)$, which obviously contradicts the former statement. In their prospect theory, Kahneman and Tversky attempt to overcome these paradoxes by postulating an even broader concept than expected utility, which subjects would again attempt to optimize.

Optimization is also a core principle outside the life sciences. In chemistry, electrons belonging to different atoms interact with and move to other atoms in such a way that they minimize energy. In physics, soap bubbles are thought to search for a surface structure to minimize energy (given the volume of air blown into them). Are economic consumers and soap bubbles really (consciously, deliberately . . .) optimizing a function? This is perhaps an interesting philosophical question, but it is not relevant for our purposes. The concept of a goal function and its optimization allows a principled approach to the study of consumers and soap bubbles, but also cognition. I will apply this concept throughout the book.

Now that you see why it makes sense to assume that humans strive toward optimizing a goal function, let's see how this can be done. First, note that optimization can mean either of two things: minimization (e.g., of energy) or maximization (e.g., of reward). In this example, I will consider minimization of the function $y = f(x) = (x-1)^2$, shown in figure 1.3. For this particular function, it is easy to find its optimum (minimum). Again, plugging in the value $x = 1$ makes the function equal to $f(x) = 0$, whereas all other values result in $f(x) > 0$. Hence, $x = 1$ is the minimum. To find the minimum in a more systematic manner, one sets the function's first derivative equal to zero and then finds the point that satisfies the resulting equation. In particular,

$$\frac{df(x)}{dx} = \frac{d(x-1)^2}{dx} = 2(x-1) = 0.$$

If the notation in this equation looks unfamiliar to you, then please check out the "Exercises" box at the end of this section, which will help you refresh (or construct) your memory.

It's obvious that this equation is satisfied when $x = 1$ (and only then). To be really precise, we should also check that we have not accidentally found a maximum because derivatives are equal to 0 at both minima and maxima. However, the quadratic nature of $f(x) = (x-1)^2$ quickly assures us that $x = 1$ must be a minimum, not a maximum (or check out the shape of the curve in figure 1.3 to convince yourself of this fact). So we found the minimum by setting the derivative of $f(x)$ to zero!

Optimization was easy in this case because we had the explicit function at our disposal and the function was very easy to differentiate. In practice, this usually will not work. When models are intended to mimic human cognition, they rarely are so simple that they can be expressed with a goal function whose optimum can be analytically found. Often, an explicit form of the function $f(.)$ to be optimized is not available; and even if it is available, it tends to be so complicated that differentiating and solving the function as I did here are very hard to accomplish. Fortunately, the field of numerical mathematics has developed a number of algorithms for finding the optimum of a function even in cases when that function is not analytically solvable. Many of these algorithms, however, require that the functions are differentiable (i.e., that a derivative can be calculated). I will also focus on such algorithms.

One of the simplest algorithms in this class is called *gradient descent* when the goal is minimization (rather than maximization). In gradient descent, one iteratively updates

What Is Cognitive Modeling?　　　　　　　　　　　　　　　　　　　**11**

the value of x (the variable to be optimized) by taking small steps in the direction of minus the derivative (= gradient). The change in x between two trials is called Δx (pronounced "delta x"), with delta (the Greek counterpart to the Latin letter d) being the first letter of *difference*. In particular, we perform the following calculation:

$$\Delta x = -\alpha \frac{df(x)}{dx}. \tag{1.1}$$

The intuition behind this rule is that when the derivative $df(x)/dx$ is equal or close to 0, then the steps Δx also will be equal or close to 0, and the algorithm has converged. The minus sign makes sure that we end up at a minimum rather than a maximum. In the current example, calculating the derivative yields the update rule

$$\Delta x = -\alpha \frac{df(x)}{dx} = -2\alpha(x-1). \tag{1.2}$$

The parameter α determines the size of the steps that we take in parameter space. When $\alpha = 0$, there is no change; small values lead to small steps, and thus slow processing—or stated otherwise, a slow time scale—occurs. This is why we call it a *scaling parameter*. Larger values of α lead to bigger steps in parameter space and thus faster processing.

The process that I just described is illustrated in figure 1.3 for the first few steps of the algorithm (x_1 to x_4). The rightmost point is the starting point (x_1), the second point (x_2) is just to its left, and so on. As we continue the algorithm, the x-values reach the optimal (minimal) point more closely.

To avoid clutter, only four points (x_1 to x_4) were shown in figure 1.3. Table 1.1 shows the process numerically for the first ten steps (x_1 to x_{10}). I used $\alpha = 0.2$, but here, as for all tables and figures in the book, you can play with the online code yourself to bolster your intuition of what is going on (e.g., see exercise 1.2). Unless all of this is already crystal clear to you, please practice in order to gain a full understanding. As you can see in table 1.1, the algorithm has almost converged after ten steps. The x-value is close to the correct value ($x = 1$), and the derivative ($df(x)/x$) and step size (Δx) are close to zero.

Table 1.1
Gradient descent on the function $y = (x-1)^2$

Step	x	y	$df(x)/dx$	$\Delta x = -\alpha\, df(x)/dx$
1	2.7	2.89	3.4	–0.68
2	2.02	1.04	2.04	–0.41
3	1.61	0.375	1.22	–0.25
4	1.37	0.049	0.73	–0.15
5	1.13	0.017	0.44	–0.09
6	1.08	0.006	0.26	–0.05
7	1.05	0.002	0.16	–0.03
8	1.03	0.001	0.10	–0.02
9	1.02	0.000	0.06	–0.01
10	1.01	0.000	0.03	–0.01

In figure 1.3, we started at an x-value larger than the optimal point (i.e., $x > 1$). In this case, we have to move "to the left" to reach the optimal point (i.e., $x = 1$). Therefore, our individual steps (Δx; see the last column of table 1.1) are always negative. If instead the algorithm starts at a value x smaller than the optimal point (i.e., $x < 1$), then plugging x into equation (1.2) shows that $\Delta x > 0$, and thus x will increase. Wherever the starting point is, the closer the algorithm gets to the optimal (minimal) point $x = 1$, the smaller the steps Δx become.

Finally, this works if one intends to minimize a function; but what about maximization? In this case, the algorithm steps in the direction of plus the derivative rather than minus. Thus, when maximizing a function, the appropriate algorithm is called *gradient ascent,* and it becomes

$$\Delta x = \alpha \frac{df(x)}{dx}.$$

Examples of gradient ascent will also be discussed here.

One of the nice things about gradient descent (or ascent) is that this algorithm works for any differentiable function (not just a quadratic one, as I used for this discussion), and for any number of variables (not just one, as I used here). A formal justification of gradient descent and related algorithms can be found in any textbook on numerical optimization (e.g., Gill et al., 1982). I will apply the concept of gradient descent throughout this book. It is therefore worth studying the equations and examples until you really understand how it works and why. I included exercises 1.2 and 1.3 to help you in this regard.

Gradient descent/ascent is just one method for finding an optimum. It just happens to be used very often because it is intuitive, is simple to apply, and makes efficient use of locally available information (i.e., near the point x) about the function to be optimized. However, it does not always apply. For example, when the goal function is not differentiable, a derivative cannot be computed, and thus gradient descent makes no sense. Other methods for optimizing a function do exist, which are not based on differentiation. Some of them will be used in this book too. For now, though, the key point is that one can formalize the cognitive process of attempting to reach a goal via optimizing a function. The gradient descent algorithm just happens to be an efficient means of doing so.

Exercises

1.1 If necessary, refresh your knowledge of calculus by consulting outside sources (e.g., xaktly .com).

1.2 In the Github repository, open the code file ch1_figs.py (i.e., code for generating the table and figures in chapter 1). Change the parameters (e.g., function, scaling parameter, number of steps), and see whether you understand how doing so changes the output of the code.

1.3 Consider the function $f(x) = A \times (x - B)^2 + C$, which we want to minimize. What influence do parameters A, B, and C have on the optimal value? Next, implement a gradient descent algorithm that minimizes the function. Try to reason it through first, and then implement the algorithm to check your answer. Compare the result of your algorithm to your analytically found optimum.

What Is Cognitive Modeling? 13

1.4* Suppose that you do *not* have the function prescription $f(x) = A \times (x - B)^2 + C$, but you can ask the computer for function values $f(x)$ at specific values of x. Can you find an algorithm to optimize the function in this case? This is actually the most typical situation that we encounter in modeling; we often cannot calculate or derive a function explicitly, but we can sample from it.

1.5* Gradient descent uses only first-order derivative information; other algorithms use second-order derivative information as well. Look up such an algorithm (e.g., on the internet) and apply it to the function $f(x)$.

1.6* Gradient descent can also be used if the function has more than one variable as an argument (in fact, in virtually all cognitive algorithms, the multivariate version of gradient descent is used). Can you generalize the gradient descent algorithm to an arbitrary number of variables? Look up your solution on the internet.

1.7 The scaling parameter α should be sufficiently small; it is typically chosen between 0 and 1, but in general the goal function determines the range of α values for which the gradient descent algorithm works well. For the function $f(x) = A \times (x - B)^2 + C$ (with $A > 0$) show that the gradient descent algorithm will not work if $\alpha > 1/A$, in which case it will overshoot and get farther from (rather than closer to) the optimum. In the code for Exercise 1.3, you can also check the effect of scaling parameter α. A more general condition on the scaling parameter (not just for quadratic functions) can be found in Bishop (1995) but is beyond the current scope.

TensorFlow

So, if modeling is all about optimizing a goal function, then in principle you need just one general optimizer package in Python for all your modeling—right? In theory, the answer is yes; and the package TensorFlow (Abadi et al., 2016) is based on just this idea. With this package, one can construct a graph whose nodes represent the operations of the model and whose edges represent the flow of information between those operations, represented in multidimensional matrices (which are called *tensors*). One can subsequently run an optimizer object that optimizes the graph toward its parameters. A detailed explanation of this package is beyond the scope of this book, but there are several good tutorials online. In the GitHub repository, you will find TensorFlow-based Python code for several chapters of this book (including this chapter), so that you can run the models discussed in that chapter (all TensorFlow code in the GitHub repository has *tf* in its name). If you don't completely understand everything stated here right now, don't worry—take a look at the tutorials, the online code, and the subsequent chapters and then return to this section some other time.

You can program cognitive models without TensorFlow, but one advantage of this package is that it helps users to think explicitly in a goal-optimizing framework. One other advantage is that it contains code for constructing some very complex state-of-the-art models, including some discussed in later chapters (especially chapters 4, 5, 8, and 9). So if you want to scale up your modeling efforts at some point, this package might be a good choice.

Minimizing Energy or Getting Groceries

A number of goals will be considered in this book, including maximizing reward (e.g., chapters 8 and 9), minimizing energy (e.g., chapters 2 and 11), and minimizing error (e.g., chapters 4 and 5). You will likely agree with the current line of reasoning that you are typically busy in daily life with attempting to obtain some goal; but you may argue with me that daily-life goals, such as serving a tennis ball, buying groceries, or passing an exam, tend to be more mundane than minimizing energy or maximizing reward. This is true, of course. However, it is important to keep in mind that the goal minimization framework is intended to be an approximation of human cognition and behavior. We do not strive to mimic cognition to the smallest detail, including a subject's introspection about human goals. Furthermore, this approximation is very often quite accurate. For example, monkeys behaving in the lab indeed are typically brought in a deprived (i.e., thirsty) state so that they will attempt to ingest as much fluid as possible. Or consider yourself playing tennis: your goal is again obviously an optimization problem; you attempt to maximize the number of games you win. You may not be immediately clear about how to obtain it, but it is definitely a maximization goal.

Moreover, it is perfectly possible to apply the optimization framework to any specific goal (such as getting groceries). The only requirement is that the goal be described as a function across a multidimensional space of adjustable parameters (Juechems & Summerfield, 2019). Consider again the farmer who has to decide how many apple trees to buy. Now, however, her gain and cost functions are a bit more complicated. Specifically, the amount of money she can expect to gain by planting x apple trees equals Ax; and the cost for planting x trees equals $Bx^2 + C$ (so the cost per tree increases with each additional tree, perhaps because the trees have to be obtained from farther distances if more trees are already purchased). The total gain to be obtained is then $Ax - Bx^2 - C$. For concreteness, let's assume that $A = 10$, $B = 1$, and $C = 2$. You can maximize $f(x) = 10x - x^2 - 2$ either analytically or numerically (via the code obtained online, as previously noted); in both cases, you will find that $x = 5$ is the best number of trees to buy. One can imagine that the farmer "climbs" this function by first buying one tree and observing the profit, then another tree and again observing the profit, and so on, until profits start to decrease. At that point, she will know that she should buy 5 trees. To be precise, using this incremental-buying algorithm, she will know only that $x = 5$ is the optimal number of trees after buying one tree too many. Still, you may note that the goals, as formulated here, drive behavior only indirectly: goal representations do not actively steer behavior. Active goal representations that drive behavior are indeed rare in modeling but definitely possible to implement. We will discuss one such model in chapter 5.

We will consider more sophisticated function-climbing algorithms (including the gradient ascent algorithm discussed before) in subsequent chapters. An interesting link to make here is with natural search algorithms from biology (Hein et al., 2016). For example, the bacterium *Escherichia coli* can sense a chemical gradient across time (while swimming), and it will continue swimming if the gradient (e.g., food) is positive, or reorient if it is negative (called *chemotaxis*). When the food gradient diminishes, it will move around in random directions and thus exploit the top of the food source (Hills, 2006). In general, single cells, whole organisms, and groups of animals search the environment via ascending positive or descending negative gradients that exist in their physical environment. Thus, the modeling framework proposed here can be seen as a generalization of a two-dimensional or three-dimensional search in physical space toward a more abstract search in a potentially high-dimensional space (Hills et al., 2012).

In the context of figure 1.1, I previously discussed spatial and temporal levels of modeling, but it is useful to link the current discussion of goals and goal functions with another conception of "levels"—namely, the famous three levels of investigation of behavior proposed by the computational neuroscientist David Marr (Marr, 1982). According to Marr, the highest level of explanation is the goal level;[2] it describes what an organism wants to do (or needs to do), given the means and environments that the organism is in. In the current framework, the goal level would correspond to the goal function that the organism has and aims to optimize. The middle level of explanation is the algorithmic level; it describes the algorithm (set of rules) that the agent uses to attain the goal. In the current framework, gradient descent (or gradient ascent) on the goal function from the highest level corresponds to the algorithm that the agent uses. Finally, the lowest level of explanation is the physical level, which specifies how the algorithm is biologically implemented. Different models make assumptions (with different degrees of biological realism) about the concrete implementation of gradient descent in actual neurons.

There will be some situations in which the optimization framework does not apply in a straightforward way. For example, some models were developed to match empirical data rather than derived from an explicit goal function. However, even in such cases, an exact or approximate goal function can often be found. And regardless of how a model was derived, the tools covered in chapters 6 and 7 (estimating and testing models) can inform the cognitive modeler of whether a particular model provides a good fit to the data or not. From a cognitive neuroscience perspective, that should be the ultimate criterion on whether a model is valid as a cognitive process.

In summary, cognitive modeling is a conceptual tool that helps building better theories across different levels of processing (spatial scale) and across different temporal scales. The models that will be considered in this book start from the assumption that cognition is typically an attempt to optimize some goal. Standard mathematical tools can then be used to implement this optimization. The process of attempting to reach the goal generates the model's dynamics. A model derived in this way provides a principled contact point for empirical data.

2 Decision Making

What I cannot create, I do not understand.
—Richard Feynman

Minimization in Activation Space

As the quotation at the top of this chapter suggests, really understanding a domain implies that you can (re-)create it yourself, so it's "in your hands," as it were—be it learning Spanish, computational modeling, cooking, or indeed any other skill. So let's get right to it. As a first attempt at cognitive modeling, let's consider decision making. Decision making is a very popular theme in modern cognitive neuroscience (e.g., Kennerley et al., 2006), but one must realize that the term as used in cognitive neuroscience is even broader than it is typically used in daily life. Specifically, decision making can apply to any situation where an agent must choose between two or more actions, from the artificial ("Is this string of letters a word or a nonword?" "Are there more dots moving left than right?") to the mundane ("Should I eat rice or pasta?") to the life-changing ("Should I study anthropology?").

Here, let's start simple and try to build a model that differentiates cats from dogs. Dividing objects or animals into distinct classes is a task that infants learn quite early in life. Thus, if we were interested in human categorization or child development, this might be our target of interest for cognitive modeling. In addition to making a model on paper and in computer code, we can go a step further and implement the model as an actual physical machine. For example, one could imagine that the machine must learn to dispense cat food to cats and dog food to dogs (see exercises 2.1 and 2.2). In fact, because the algorithms we use in modeling are also used to train machines, the domain is sometimes called *machine learning* (Murphy, 2012). However, in machine learning, the focus of interest is on building algorithms or devices that perform practical tasks (or at least tasks that bring profit to their creator): Think of self-driving cars by companies like Google and Uber, which are trained based on machine learning principles. Here, however, we are not interested in building machines, but rather on understanding the human mind. I will therefore not use the term "machine learning."

To differentiate cats from dogs, let's further assume that we have a set of feature detectors, each of which responds when some feature in the environment is present. Feature detectors could detect low-level perceptual properties like stimulus orientation, size, and luminance. But let's consider three more high-level detectors. One detector finds whether the animal has its picture on Facebook; a second checks whether it has four legs; and a third investigates whether the animal bites visitors (see figure 2.1a). Don't ask how these detectors can do that. But they can, and based on these three detectors, we attempt to build our machine. Feature detectors are either off (represented by 0) or on (represented by 1). Alternatively, feature detectors could take on some activation value x, in the range 0 to 1. In the latter case, you can think of activation as the probability that the detector considers the feature to be present. For example, a value of $x = 0.8$ for the "bites visitors" feature would mean that the machine is pretty (but not completely) sure that the current animal would bite a visitor when the visitor dares to come too close.

Our model also has a cat detector, which receives the activation of the first three detectors as input (for the Facebook, four legs, and biting features, respectively). This cat detector has an activation level, x_{cat}. One can think of this activation as identifying the probability of there being a cat in the environment (or at least, in the cat detector's receptive field). The detector is like a nerve cell (or neuron) that attempts to respond

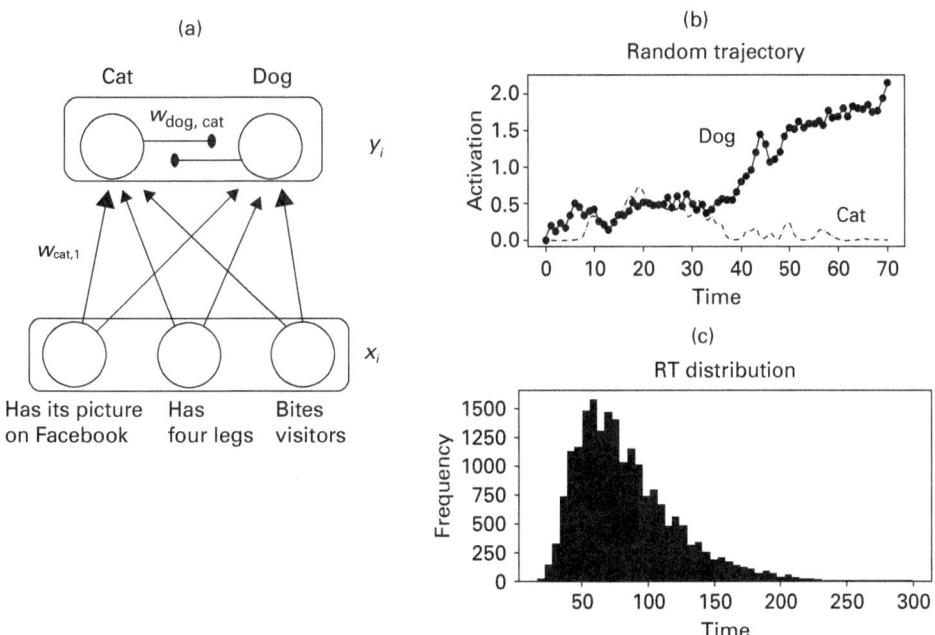

Figure 2.1
(a) Pet detector. To avoid clutter, only two of the weights are labeled. (b) Two trajectories of a cat and dog unit for a particular (doglike) input activation pattern. (c) RT histogram.

Decision Making   **19**

when (and only when) a cat appears in the environment. Similarly, we have a dog detector, which has an activation level x_{dog}; it is active when a dog is present.

Exercises

2.1 What other features could feed into the "cats versus dogs" categorization machine?

2.2 Think of another context where a machine separating two other types of objects (or organisms) could be useful. Which features could it use?

The next step is: How does one go from the feature detectors to the pet detectors? A key assumption is that the activation levels of the feature detectors are combined in a linear way and then sent to the next stage for further processing (here, pet detection). This is the important *linear principle*, which will be used over and over again in this book and in cognitive modeling more generally. To make this more concrete, suppose that the three feature detectors have activation values x_1, x_2, and x_3; Then the input to each of the pet detectors is as follows:

$$in_{\text{cat}} = w_{\text{cat},1}x_1 + w_{\text{cat},2}x_2 + w_{\text{cat},3}x_3. \tag{2.1}$$

$$in_{\text{dog}} = w_{\text{dog},1}x_1 + w_{\text{dog},2}x_2 + w_{\text{dog},3}x_3. \tag{2.2}$$

Equations (2.1) and (2.2) hold that the input to each detector (e.g., in_{cat} to the cat detector) is a linear combination of the input (x_i) that it receives from three other units. The linear transmission coefficients w_{ij} are called *connections* or *weights* w_{ij} from detector j to detector i. A weight from detector j to detector i is thus represented as w_{ij}, consistent with the (slightly confusing) tradition of putting the output unit in the first index and the input unit in the second index of weight w. In figures, positive (or excitatory) weights are typically represented with an arrowhead ending (like $w_{\text{cat},1}$ in figure 2.1a), while negative (or inhibitory) weights are typically represented with a dot ending (like $w_{\text{dog,cat}}$ in figure 2.1a; more on this parameter will be discussed in the next section).

In principle, one could combine the features in other, nonlinear ways. However, a similar story holds as for the optimization principle. First, linearity is a plausible principle, in that it implements the basic architecture of a synaptic junction (figure 2.2) (McLeod et al., 1998). In particular, one can think of activation values x_i as neural firing rates, and of the weights w_{ij} as the neural connection strengths from neuron j to neuron i. Second, linearity is a mathematically very well developed principle. In fact, linearity is so well studied and well understood that a whole branch of mathematics is devoted to how linear combinations of variables behave (linear algebra). Also outside the field of linear algebra in the strict sense, linearity crops up everywhere in science and mathematics. This is not to say that all models obey only linear principles. In fact, there are several nonlinear models, which have several interesting properties that originate from nonlinearity. However, "nonlinear model" is an ambiguous and misleading expression because a nonlinear model is typically built on a linear core that does much of the model's heavy lifting. If you're studying models without any linearity in them whatsoever, you're in a pretty exotic area of model space! Because the linear principle is used throughout this book, and in modeling more generally, it is essential to understand equations (2.1) and (2.2) before continuing. Exercises (2.3) and (2.4) will help in this regard.

20   Chapter 2

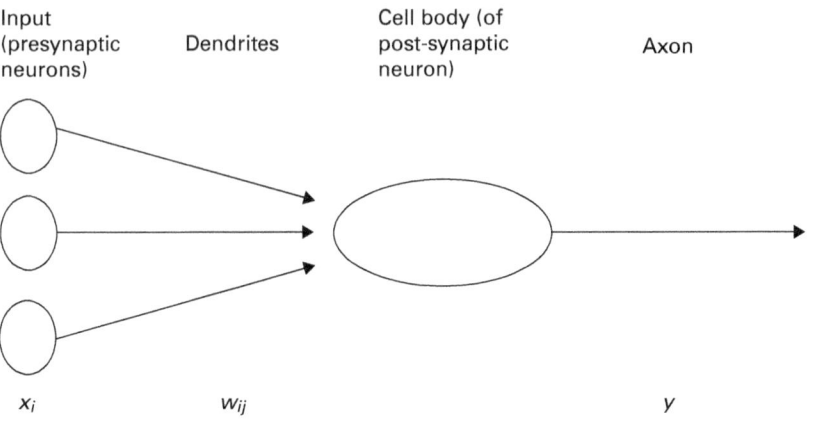

Figure 2.2
The linearity principle in a synaptic junction.

Exercises

2.3 Take the code ch2_linear_model.py, and study what happens with different input values
for the input units.

2.4 In the same code, study what happens when you increase or decrease the weights w_{ij} in
equation (2.1).

Recall that the three features at our disposal for distinguishing cats from dogs are "has
its picture on Facebook," "has four legs," and "bites visitors." Features 1 and 3 are plausibly more connected to cats and dogs, respectively; accordingly, weights $w_{cat,1}$ and $w_{dog,3}$
are large, whereas weights $w_{cat,3}$ and $w_{dog,1}$ are small. In contrast, feature 2 doesn't clearly
distinguish between cats and dogs, so weights $w_{cat,2}$ and $w_{dog,2}$ are of approximately the
same size. The weights w_{ij} are treated as fixed in this chapter, but in several subsequent
chapters (including chapters 3–5), I will describe methods to change these weights.

Equations (2.1) and (2.2) already embody a minimal model of cognitive pet processing. Indeed, several models are based on nothing but linear transmission of information across detectors; or at least, linearity plays a large role in several interesting and
influential models (e.g., Rescorla & Wagner, 1972). However, to make the pet detector
model more cognitively interesting, in the next section, I will define a goal that the
model can strive toward. I can then apply the optimization principle and thus make
sure that the model attempts to reach that goal.

Is the pet detector model a biologically plausible model of how humans detect dogs
and cats? As discussed in chapter 1, one key lesson about modeling is that the answer to
this question depends on the domain that one is interested in modeling. If one wants
to model the behavior of single neurons in the primary visual cortex, it makes little
sense to have neurons with binary (0/1) activation profiles. A biological neuron exhibits much more complex behavior, so if the level of interest is the single neuron, a binary
neuron is entirely inadequate. In contrast, when modeling a higher-level domain, such
as categorization of pets, one may be interested in more high-level cognitive questions,

Decision Making   21

such as whether the model can extract a prototype dog or cat; whether the progression of knowledge as the model develops is similar to the progression that human children make; whether one can make predictions for functional magnetic resonance imaging (fMRI) studies in pet categorization; and so on. In this case, we may not need very high levels of biological realism, and a simple binary neuron that responds to the presence of cats (or dogs) may be enough. In fact, as discussed in chapter 5, a very extended version of this "cats and dogs" model explains several properties of human categorization (Rogers & McClelland, 2004).

A Minimal Energy Model

So, the cat and dog detectors (figure 2.1a) receive a linear combination of the feature detectors, as specified in equations (2.1) and (2.2). What should these detectors do with this information? Let's assume that the detectors are mutually connected by negative (inhibitory) weights $w_{\text{dog,cat}}$ (from cat to dog, as labeled in figure 2.1a), and $w_{\text{cat,dog}}$ (from dog to cat). A key question concerns the dynamics of the model—that is, how the activation values of its detectors change across time. These dynamics are crucial for cognitive neuroscience because they allow one to make predictions about accuracy, response time (RT), neural activation, and other dependent variables of interest. To define the model dynamics, one first defines a goal function (as discussed in chapter 1), here called an *energy function*. One then assumes that the organism aims to minimize this function; the path toward this minimization will determine its dynamics. If this sounds a bit abstract, don't worry—just keep reading. It will become clear with the concrete examples that appear later in this chapter. So, in this case, let's define the energy function E as follows:

$$E = -in_{\text{cat}}y_{\text{cat}} - in_{\text{dog}}y_{\text{dog}} - wy_{\text{cat}}y_{\text{dog}}, \tag{2.3}$$

with the parameter $w < 0$. In this specific case, I consider energy function E as a function of the activation y_{cat} and y_{dog} of the pet detectors, so I could have written E as $E(y_{\text{cat}}, y_{\text{dog}})$ to get a more complete (but also more burdensome) notation. Note that I use the letter y for representing the activation of the cat and dog detectors. Whenever there is a clear directionality between layers of neural units (as in figure 2.1a, where activation flows from features to pet detectors), I use consecutive letters for the activation variables $(x, y, \ldots)$. If there is no clear directionality and each unit can in principle activate all units, I use a single letter such as x to indicate detector activation.

Don't worry where this energy function comes from. Let's just assume that the pet detector's goal is to minimize it. In other words, assume that the pet detector changes activation $y_{\text{cat}}, y_{\text{dog}}$ in such a way to make energy E as low as possible. Note that the energy can also be negative. You may find this a bit disturbing, but again, don't worry about it; just consider E as a function that one strives to make as low as possible. Finally, note that the activation values y_{cat} and y_{dog} take on continuous values. If they were binary-valued, like the input detectors x_i, gradient descent in their activation space would make no sense. Can you see why?

Let's try to get an intuitive grip on what minimizing energy function E means. First, let's assume that the input to the cat detector, in_{cat}, is high. Then, it would be advantageous to make y_{cat} large as well. Indeed, if both in_{cat} and y_{cat} are large positive values, inspection of equation (2.3) should convince you that E will become more negative. Furthermore, inspection of the same equation shows that because w is negative, and because the system attempts to minimize E, it attempts to avoid states where both y_{cat} and y_{dog} are high; such states would increase energy. Hence, the negative weight w incorporates the assumption that the states are unlikely to be active together. Stated in another way, they inhibit one another; if one is active, it will inhibit the other to make sure that the other does not become active as well. Altogether, then, large values of in_{cat} should make y_{cat} large; large values of in_{dog} should make y_{dog} large; but y_{cat} and y_{dog} should not be active simultaneously. That's the intuition behind minimizing E. Do exercise 2.5 to further prime your intuition.

Now that the intuition is clear, the next issue on the agenda is how to minimize E in a systematic manner. Recall from chapter 1 [equation (1.1)] that the gradient descent rule stipulates to descend a goal function (f in chapter 1; E in chapter 2) along minus the derivative of that function. This process will typically require many iterations, or steps. The steps needed to find this minimum allow one to model RT. Taking many steps corresponds to a slow RT; taking few steps corresponds to a fast RT. The final state in which the model (e.g., the pet detector) ends can be used to model the choice that it has made (e.g., is there a cat or a dog in the visual periphery?). In this way, one can model choice and RT data simultaneously. This is exactly the rationale behind the popular class of sequential sampling models of decision making, which includes several models such as the linear ballistic accumulator model (Brown & Heathcote, 2008), or the diffusion model (see the discussion later in this chapter and in Ratcliff, 1978). It is a very useful property for cognitive neuroscience, given that most empirical data sets contain at least choices and RTs. These variables constitute two major targets for behavioral modeling. Other variables that have been modeled with this framework include trial-by-trial confidence judgments (Zylberberg et al., 2016) and neural variables derived from electroencephalography (EEG) and fMRI (Mansfield et al., 2011).

I next consider how to take steps in activation space that are appropriate for the purpose of minimizing the energy function E. In particular, by applying the gradient descent principle to the energy function E, one obtains the following recipe for finding the minimum of the energy function for the cat detector activation y_{cat}:

$$\Delta y_{cat} = -\alpha \frac{dE}{dy_{cat}} = \alpha \left(in_{cat} + w y_{dog} \right),$$

and similarly for y_{dog}:

$$\Delta y_{dog} = -\alpha \frac{dE}{dy_{dog}} = \alpha \left(in_{dog} + w y_{cat} \right).$$

I will discretize time and assume that it occurs in steps that are indexed $t = 1, 2, \ldots$ Such steps may correspond to trials in a behavioral experiment, stimulus presentations to an

Decision Making   **23**

observer, or any other appropriate unit of time. When necessary to avoid ambiguity, I will write activation (e.g., x) as an explicit function of t [e.g., $x(t)$]. Rewriting this equation, it follows that one should change y_{cat} in the following way from time step $t-1$ to t:

$$y_{\text{cat}}(t) = y_{\text{cat}}(t-1) + \alpha\left(in_{\text{cat}} + wy_{\text{dog}}(t-1)\right)$$

However, for the purpose of modeling cognition, this equation has a disadvantage. Repeating the process would always yield the exact same trajectory $y_{\text{cat}}(t)$ if the input in_{cat} remains the same. This issue is solved if one assumes that there is random noise in the process. This is hardly a strong assumption: Indeed, it can be assumed that our brain is slightly noisy, and some noise invades each cognitive process. Formally, the assumption here is that at each time step t, a noise variable $N(t)$ with a normal distribution with mean 0 and standard deviation σ is added to the main y equation for cats and dogs as shown in equations (2.4) and (2.5), respectively:

$$y_{\text{cat}}(t) = y_{\text{cat}}(t-1) + \alpha\left(in_{\text{cat}} + wy_{\text{dog}}(t-1)\right) + N(t) \tag{2.4}$$

and

$$y_{\text{dog}}(t) = y_{\text{dog}}(t-1) + \alpha\left(in_{\text{dog}} + wy_{\text{cat}}(t-1)\right) + N(t). \tag{2.5}$$

It is important to note now that the updated equations (2.4) and (2.5) depend on the parameters w_{ij} that were used before, as illustrated in figure 2.1a. First, the equations depend on in_{cat} and in_{dog}, quantities, which themselves depend on parameters w_{ij}, as shown in equations (2.1) and (2.2). Second, note that the terms wy_{dog} and wy_{cat} in equations (2.4) and (2.5) respectively, represent the extent to which the dog detector competes with the cat detector. The parameter w from the energy function, therefore, can be thought of as representing the inhibitory interaction between cat and dog detectors, and specifically that this inhibition is symmetric, or $w_{\text{cat,dog}} = w_{\text{dog,cat}} = w < 0$. It would also be possible to have asymmetric inhibition: For example, the cat detector might compete against its neighbors more strongly than the dog detector does (i.e., $w_{\text{cat,dog}} > w_{\text{dog,cat}}$). In such case, the energy (and thus update) equations would become a bit more complicated.

Once we have parameters w_{ij} and an energy function (and thus, the update rule), we can run the model and evaluate its performance. Even though this minimal energy model is quite simple, it already exhibits some interesting properties. Figure 2.1b shows the trajectories of the cat and dog detectors in one trial (one trace for each of the two detectors). Figure 2.1c shows the RT distribution of the model across 20,000 replications (simulated with ch2_figs.py); the distribution is smoother with more replications (see exercise 2.7). It is reassuring that the model shows positive skew, which is a typical feature of any empirical RT distribution (Ratcliff, 1978). Of course, positive skew is only one feature of empirical data that we want to model.

A second interesting feature of the model is that it shows a distance effect. In particular, if the animals resemble each other, then evidence for the cat and dog detectors becomes similar (i.e., in_{dog} becomes similar to in_{cat}). As a result, RTs become slower and accuracy decreases. This is known as the *distance effect*. In psychophysics, the distance effect has been well known to apply for continuous variables (e.g., size, weight, . . .) for more than a century; for example, it is easier to judge which of two weights is heavier if they are more different (distant). The distance effect is formalized in psychophysics in the famous Weber law (jointly with the size effect, which holds that discriminating between two large quantities is harder than discrimination between two small quantities). The distance effect was also reported for symbols expressing abstract variables in Arabic numbers. This distance effect in Arabic numbers has been interpreted as evidence for a spatial-Euclidean representation of numbers (Dehaene, 1997; Moyer & Landauer, 1967). However, the fact that, as seen in our cats and dogs example, *any* quantitative difference between in_{cat} and in_{dog} gives rise to a distance effect, shows that it is very hard to draw strong conclusions about the spatial nature of representations from just the distance effect (Verguts et al., 2005). This example illustrates one advantage of cognitive modeling: It tells us which inferences from data may or may not be justified. This, of course, does not imply that numbers are *not* spatially represented—just that the distance effect is uninformative in this respect.

Exercises

2.5 Make a plot of the energy function [equation (2.3)] as a function of y_{cat} and y_{dog}. Investigate whether it has the right qualitative properties by checking the effect of in_{cat}, in_{dog}, and w on the shape of the function. For example, if $in_{cat} > in_{dog}$, the energy function should encourage activation of the cat output unit, but inactivation of the dog output unit. Also, if in_{cat} is closer to in_{dog}, it should take longer to reach the optimal (low-energy) point (this is the distance effect). You can use ch2_plot_energy_fun.py on GitHub for this purpose.

2.6 Derive equations (2.4) and (2.5) from the gradient descent rule described in chapter 1.

2.7 Using ch2_figs.py, run some simulations to see that the RT distribution becomes smoother with more trials. Convince yourself that positive skew is a natural property of this model. Change the parameters and predict how these changes will influence the final results. Finally, check your predictions using Python.

2.8 Adapt the code in ch2_figs.py to check whether the "cats and dogs" model shows a distance effect. Can you think of a model that does *not* exhibit a distance effect?

Once one has a basic model, one can choose to extend it based on theoretical or practical reasons. One useful extension is to rectify the model activations, meaning that activation of its detectors never goes below zero. This extension is not necessary, but if we intend to interpret the model as a neural model, it is a plausible move. The two trajectories in figure 2.1b are actually the result of cat and dog detectors in a rectified model. An alternative extension could be to add a decay parameter to the model (Usher & McClelland, 2001). In this case, when there is no input, the activation has the tendency to decay toward zero. This has the useful computational property that it keeps activation on a leash; activation will not drift off to infinity if a leakage term is attached to its units. With a decay parameter, there is typically an upper bound that activation of the model units tends toward but does not reach. In this sense, adding a decay term can be considered as a soft version of rectification at an upper boundary.

In general, modelers typically start with some goal that the model must achieve (e.g., minimize energy), but then they can add parameters as desired. Whether specific parameters are justified for computational reasons, empirical reasons, or just for aesthetics is to be determined by a modeler's peers, just as in other instances of scientific theory development.

Cooperative and Competitive Interactions in Visual Word Recognition

Because response (cat, dog) units interact (here, they inhibit each other), models like those developed in the previous section are sometimes called *interactive activation models*. They have been extremely influential in cognitive neuroscience, such as in visual (McClelland & Rumelhart, 1981) and auditory (McClelland & Elman, 1986) word recognition. One particularly influential one has been the *interactive activation model* (IAM) of McClelland and Rumelhart (1981), to which I turn next.

The original motivation for the IAM was to achieve the word superiority effect in the letter identification task (Reicher, 1969). This task requires the subject to detect a letter at some position. The word superiority effect entails that detecting a letter is easier in a word context than in a nonword context. For example; if subjects must identify the first letter of "SHIP," they can do so more easily (require less viewing time) than in nonword contexts (e.g., SXXX). Since the original basic finding, several extensions have been reported. For example, the advantage is also found for pronounceable nonwords (e.g., "SHIK"). Furthermore, there is no advantage of a highly constraining context relative to a weakly constraining one. For example, context _HIP (allowing 3 words) has no advantage over a context like _INK (allowing more than 10 words). See McClelland and Rumelhart (1981) for a more elaborate overview. For now, note that it's definitely not trivial to account for the word superiority effect, including all its variants. What type of cognitive architecture would generate all these (often subtle) effects, but not produce spurious (i.e., unobserved) effects? Enter the IAM.

A picture of some of the units and connections in the IAM is shown in figure 2.3. Each of the units in this model has an activation function very similar to the one from the minimal energy "cats and dogs" model. The lowest level contains units that respond to features that individual letters may have (horizontal, vertical, and diagonal lines at each possible position). This feature layer feeds into a letter layer, where each letter unit has a positive connection with the detectors for its features. Next, this letter layer projects to a word layer. Here, each unit corresponds to a specific word, so a detector for, say, "ABLE" is connected to units for "A in letter position 1," "B in letter position 2," and "L in letter position 3." Hence, because the model responds only to four-letter words, it has 26×4 letter detectors (26 for each position).

Whenever a feature pattern is presented at the feature level, activation feeds into the letter level, which uses an activation rule very similar to the one described previously. This letter layer then feeds into a word layer, which again uses a similar activation rule.

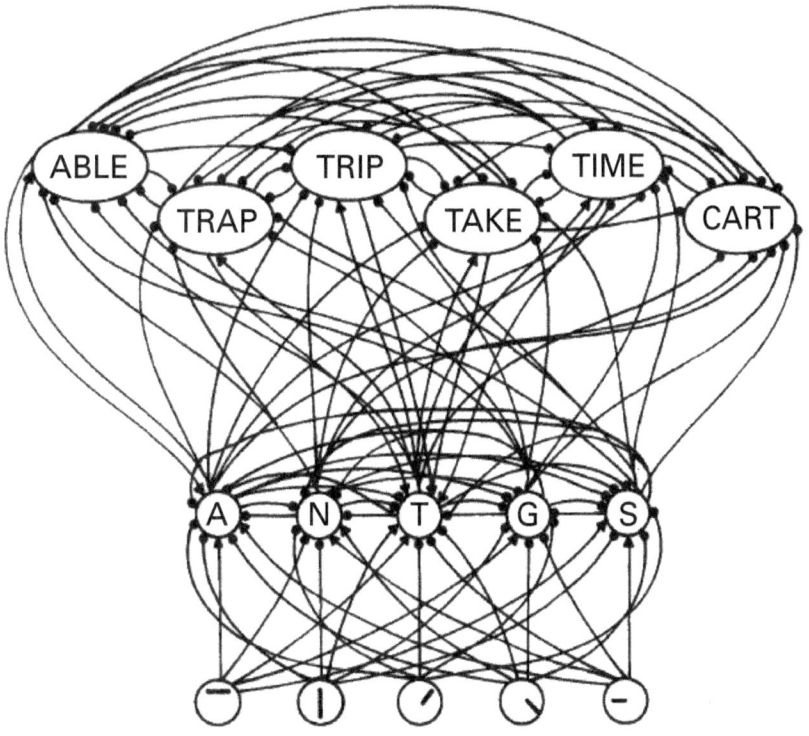

Figure 2.3
Part of the IAM of letter perception. This figure was reproduced with permission from McClelland and Rumelhart (1981).

Of note, the word layer also projects to the letter layer; hence, the letter layer receives both bottom-up (feature) and top-down (word) input. Thus, whenever a feature pattern is presented, all units are activated to varying extents. For some units, activation keeps increasing as time goes on. One can then place a threshold on the units in the word level, and whenever a word unit crosses that threshold, that word is chosen. Thus, we can model both word choice (accuracy) and word RT. Similarly, we can place a threshold on the units at the letter level. Whenever a letter unit crosses that threshold, this can be said to correspond to a decision of the model: it chooses a specific letter. Thus, we can model the letter identification task and test the model's ability to model the word superiority effect.

It is perhaps unsurprising that the IAM exhibits a word superiority effect. The reason is that words provide extra input to its constituent letters (above the feature input that those letters receive), thus turbo-charging the activation of these letters. It is less obvious that the model also accounts for all other effects reported in this context; this requires actually simulating the model. This is exactly what McClelland and Rumelhart (1981) did: they found that the model indeed exhibits several key properties. For example, it shows a pronounceable pseudoword superiority effect (letter strings like "SHIX") because such strings tend to activate several words, leading to similar effects as actual words. However, the model shows no advantage for nonpronounceable nonwords (letter strings like "SXXX"); this is as it should be, because human subjects also do not show that effect. In the model, the reason is that strings like "SXXX" tend not to activate actual words at the word layer. I again refer to McClelland and Rumelhart (1981) for a full report of how and why the model shows these effects. A good model leads to novel predictions, and the IAM indeed leads to several ones. Some of these predictions were tested in a follow-up paper by the same authors (Rumelhart & McClelland, 1982).

Later authors questioned several of the IAM's assumptions, with particular emphasis on the letter coding scheme. A vast psycholinguistic literature developed on how exactly letters (and subsequently, words) are coded in the human brain (Dandurand et al., 2010; Davis, 2010; Plaut et al., 1996). In this discussion, modeling plays a key role to back up claims that some coding scheme is or is not computationally advantageous, or consistent with neural (Vinckier et al., 2007) or behavioral (Grainger, 2008) data. More generally, the basic assumptions of the IAM (such as parallel and competitive processing) have been and still remain an active area of research in psycholinguistics (Coltheart et al., 2001; Chen & Mirman, 2012; Ness & Meltzer-Asscher, 2021).

The Hopfield Model

Different important models are based on the concept of energy minimization. Consider the network layout of figure 2.4. Detectors (or units) are considered to be active ($x = 1$) or inactive ($x = 0$). As before, an "active" detector means that the feature that the detector is supposed to detect exists in the environment (or at least, that the model thinks that it does). For example, for the detector labeled John, its activation could mean, "The person is named John." Three detectors are positively connected—namely, those representing the set $J = \{$John, male, poor$\}$. (For simplicity, if two units are mutually connected, the two arrows are sometimes replaced with a two-sided arrow, as in figure 2.4.) Three other detectors are positively connected as well, those representing the set $M = \{$Mary, female, rich$\}$. Also, as in the minimal energy model, a negative connection means that the model will attempt to avoid both connected units being active at the same time, and a positive connection will encourage the model to make both connected units simultaneously active. Such a positive connection, therefore, can be interpreted as, "If one of these features is present, then the other is likely to be so as well." Because this interpretation is symmetric, connections are symmetric as well, and one needs to define only one of w_{ij} and w_{ji} (e.g., choose w_{12}, and then fix $w_{21} = w_{12}$). Elements of sets J and M have mutually inhibitory (negative) connections. Because all units are at the same level in this case (there is no clear directionality of activation flow from one set of units to another), I use only the symbol x to indicate unit activation.

Given an appropriate energy function (with consequent activation function), this architecture forms the basis of the Hopfield model, named after the American physicist John Hopfield who developed it (Hopfield, 1982). I follow the discussion of the Hopfield model by Ackley and colleagues (1985). Let's first look at the model's activation

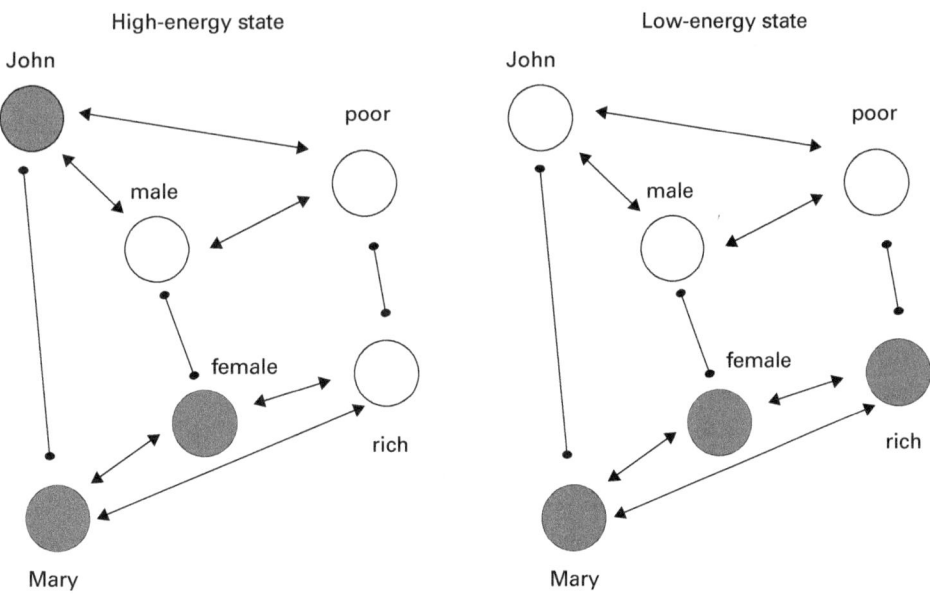

Figure 2.4
The Hopfield model; active units are indicated by gray shading.

dynamics. As in the minimal energy model, the goal of the Hopfield model is to minimize energy. Its energy function is as follows:

$$E = -\sum_{j}\sum_{i<j} w_{i,j}x_ix_j + \sum_i \theta_i x_i. \tag{2.6}$$

The $i < j$ below the summation sign means that one only considers pairs (i, j) where $i < j$. For example, one can take the term $w_{12}x_1x_2$, but it is not necessary to also include the term $w_{21}x_2x_1$. Indeed, the weights are symmetric ($w_{ij} = w_{ji}$), and thus the term $w_{21}x_2x_1$ would give exactly the same contribution as $w_{12}x_1x_2$. Obviously, I could just as well have chosen to take only pairs where $i > j$ instead of only the pairs with $i < j$. Note that equation (2.6) is very similar to our definition of energy in the minimal model: There are terms where the variables appear in a second-degree form (x_ix_j), and terms where the variables appear in a first-degree form (x_i). However, the parameters multiplying the first-degree terms x_i [the parameters θ_i in equation (2.6), but in$_{\text{cat}}$ or in$_{\text{dog}}$ in equation (2.3)] have a different sign than before [i.e., positive in equation (2.6), but negative in equation (2.3)].

Because activation is now binary (i.e., activation x is 0 or 1), we cannot use the gradient descent rule over the energy surface. Gradient descent is possible only for continuous variables. But we can still minimize the energy function, based on the following argument. Take an arbitrary unit k ($k = 1, \ldots, 6$ in the example shown in figure 2.4). Let the energy of the full system with unit k clamped (fixed) at 1 and 0 be E_1 and E_0, respectively. Then the difference between these values can be written as

$$E_0 - E_1 = \sum_i w_{i,k}x_i - \theta_k. \tag{2.7}$$

Decision Making   **29**

This leads to the following intuitive update rule: First, calculate the activation flowing into detector k (i.e., the weighted sum of activation of other units; again, look at figure 2.4). Second, check if this activation is larger than a threshold value θ_k. If it is larger, then set the unit k to 1 (because then $E_1 < E_0$, implying that E_1 is an energetically better state); otherwise, set it to 0 (because then $E_0 < E_1$).

Repeated application of this update rule will lead to an activation pattern $(x_1, \ldots, x_6)$ that is a local minimum of the energy function. Such a local minimum corresponds to a memory stored in the weights w_{ij}. Memories can thus be incorporated into the Hopfield network as local minima of its corresponding energy function, and the Hopfield model can be used as a model for memory. Depending on the weights w_{ij}, the model stores different memories.

The beauty of this approach is that it globally solves all hard and soft constraints implicitly defined by the weights w_{ij}, using an update rule that only uses local information [i.e., information that each unit has access to, as incorporated in equation (2.7)]. A *hard constraint* on a rule is a constraint that must be satisfied; not satisfying the constraint means breaking the rule. For example, in the rules of chess, a hard constraint would be to that rooks can move only in straight (horizontal or vertical) lines; or that if the king is put in danger (check), the danger must be removed on the next move or else the game is over (checkmate). A soft constraint is one that is not dictated by the rules, but that you'd better try to obey nevertheless. For example, a soft constraint is that the queen should not be endangered. It is perfectly allowed by the rules, but, in Hopfield terminology, an endangered-queen board position would have high energy and should best be avoided. However, it remains a soft constraint and can thus be overruled: if endangering your queen has consequences that compensate the bad position (e.g., because you can thereafter checkmate the opponent), it is (energetically speaking) worth doing so.

The two states (John, male, poor) and (Mary, female, rich) are called *attractors* for the model. That is, the model will be attracted to either one state or the other. Each appears as a local minimum in the energy function. Consider figure 2.4. If one starts from a state close to (John, male, poor), one will end up in that state after a few iterations of the update rule. In contrast, if one starts close to (Mary, female, rich), this is what the model will be attracted to.

We can try to visualize this attraction process. Given that there are seven units, the energy function E would be optimized in this case over a seven-dimensional space; combined with the energy function itself, one would require an eight-dimensional space to visualize the full energy landscape of this model. This is obviously hard to plot, so let's reduce the seven-dimensional input space to a single dimension. This dimension can then be represented on the x-axis, and the energy of each point on the y-axis. Such an energy function is depicted in figure 2.5. Here, there are two minima, and thus two attractors. One could consider that one of them corresponds to (John, male, poor) and the other one to (Mary, female, rich). Depending on where the energy minimization process starts, one ends up in the first (John) or the second (Mary) minimum.

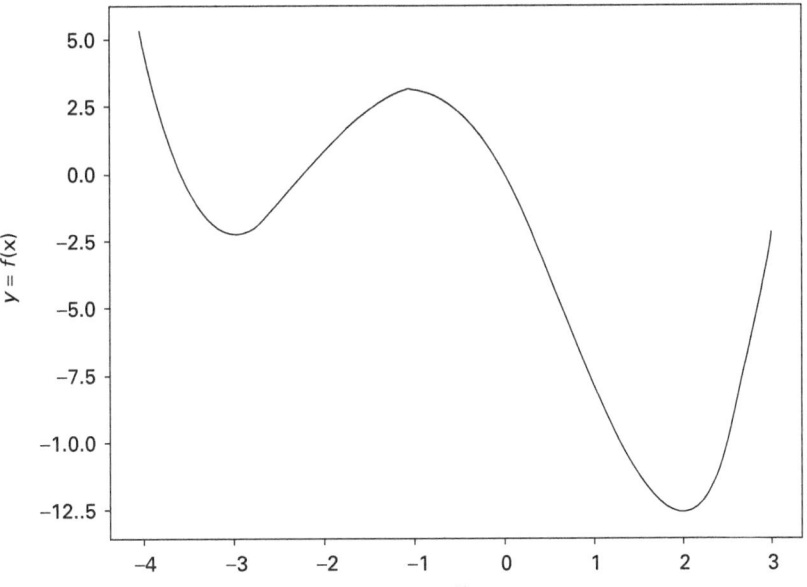

Figure 2.5
Energy function with two local minima.

Harmony Theory

The fact that the Hopfield model can deal with both hard and soft constraints is an interesting psychological property, as humans are also particularly good at doing this naturally. Think again about the chess example, where several constraints must be integrated. In realistic cases, it is nearly impossible to set up an explicit rule system that satisfies all constraints optimally ("If you play chess, do not endanger your queen, except if. . . ."). But with appropriate weights w_{ij}, the Hopfield model can end up at an optimal solution. In humans, such constraints are typically naturally integrated, usually without much conscious thinking by the agent (chess player).

Such unconscious constraint satisfaction has also been proposed to occur in judgment of grammaticality of a sentence, in both production and comprehension (Prince & Smolensky, 1997). For example, we can immediately decide that "John loves her" is grammatically correct but "John her loves" is not. How do we do that? According to harmony theory, each language grammar embodies a set of constraints via network weights; in judging the grammaticality of a sentence, the cognitive system would compute the harmony of that sentence (corresponding to negative of energy). If harmony is maximized (i.e., energy is minimized), the string is considered grammatical. The authors argue that these low-level soft constraints give rise at a higher level to hard grammatical constraints; just as lower-level interactions between molecules give rise to the higher-level concept of temperature in physics.

Decision Making   **31**

Exercises

2.9   Write the energy equation [equation (2.6)] in full (i.e., without the summation signs) for a network with 3 units.

2.10 Check that $E_0 - E_1$ has the form shown in equation (2.7).

2.11 Apply the Hopfield model activation rule using ch2_hopfield_model.py. Observe that the model attempts to find a local minimum in energy, corresponding either to John or to Mary, with their respective properties. Change the weights and observe how this changes the final activation pattern that the model is attracted to. How would you distinguish hard from soft constraints in this model?

2.12 Store a third person in the model (e.g., {Mike, male, rich}). Obviously, you will have to add at least one extra node (for Mike).

Solving Puzzles with the Hopfield Model

This natural combination of hard and soft constraints can also be used to solve computational problems. For example, the eight-rooks problem asks us to place eight rooks on a chessboard in such a way that no row or column contains two rooks. (Note: there is more than one solution to this problem.) To solve this problem, one can try several combinations using an actual chessboard. Alternatively, one can construct a Hopfield model that solves the problem. For this purpose, construct a model where each unit represents one of the chessboard positions (i.e., a model with 64 units). One can then define an energy function that incorporates the constraints of the eight-rooks problem. That is, just as *John* had to be associated with *male* but not *female* in the earlier example, we can incorporate in the energy function the constraint that two rooks cannot appear in the same row. Placing two rooks in the same row should increase energy. By constructing an appropriate energy function, one obtains a Hopfield model that will attempt to minimize energy. If the energy at the final solution is sufficiently low, it will have found a solution satisfying all the constraints of the eight-rooks problem. Rojas (1996) describes how you can construct the energy function for this problem.

Another famous puzzle that the Hopfield model can address is the traveling salesperson problem. Here, the problem is to find the shortest path between several cities (say, five) that a salesperson must take to minimize his traveling distance. Note again that finding the solution with minimal distance using an explicit rule system ("Go first from Brussels to Antwerp, except if you start in Ghent, in which case, . . .") would be nearly impossible. Instead, to solve this problem, one can construct a model with 5 units for each city (1 unit for each potential rank order in which a city is visited). The weights between all the units for each pair of cities can be chosen to correspond to the distance between the two cities. In this way, an energy function can be defined whose minimum will correspond to a minimal traveling distance. Again, running the corresponding Hopfield model will solve the problem of finding the minimal distance (although there can be several local minima in this problem). Again, refer to Rojas (1996) for a precise definition of the corresponding energy function and further information about other computational problems that can be solved with a Hopfield model.

Human Memory and the Hopfield Model

So what does the Hopfield model tell us about human cognition? Ever since patient H. M. was studied by Milner and Scoville (1957), it is known that the hippocampus is a key anatomical structure for human memory. Famously, this patient with bilateral hippocampal removal could interact normally with other patients and hospital personnel, but after a few minutes, he would have forgotten each interaction completely. Experiments with several species have replicated this basic finding several times: Hippocampal removal spares memory for facts[1] and procedures but completely impairs short-term memory (Morris et al., 1982). It remains to be determined, however, exactly *how* the hippocampus contributes to memory. In this regard, computational modeling has been extremely helpful for spelling out concrete and testable ideas on how hippocampus and cortex interact to produce memory. A key computational idea is that hippocampus functions like a Hopfield memory. This hypothesis was partly inspired by the peculiar anatomical connectivity pattern of the hippocampus and its input and output pathways. In particular, the cortex projects to the entorhinal cortex, which projects to the dentate gyrus, which then projects to CA3 (Hasselmo & McClelland, 1999; O'Reilly & Norman, 2002). Region CA3 of the hippocampus has a very strong recurrent connectivity, much more so than other cortical and subcortical regions. Such a connectivity pattern is exactly what is needed in order for a region to implement a Hopfield model. From CA3, information then flows out of the hippocampus via region CA1.

The computational argument that the hippocampus functions as a short-term memory Hopfield model goes along the following lines: Novel episodes consist of spatio-temporal patterns of activity. For example, an episode describing my breakfast today could consist of the set E1 = {"this morning," "at home," "cornflakes," "coffee," "with my kids," . . .}. This would then constitute one attractor in the Hopfield/hippocampus model. The episode describing my commute to work could consist of the elements E2 = {"raining," "by train," "delayed," . . .}. The process whereby a pattern (like E1 or E2) is encoded in memory to remain distinct from other patterns is called *pattern separation*; indeed, E1, for example, is thus separated in memory from highly similar patterns like E = {"yesterday morning," "at home," "banana," "coffee," "with my kids," . . .}.

The nice thing about the Hopfield network is that one can present a partial pattern of an attractor, and the Hopfield model will complete it toward the full attractor. Such pattern completion is an essential function of the Hopfield model. This is an interesting psychological property, as human memory functions in this way too. For example, presenting just a part of E1, say {"this morning," "at home"}, can bring back vivid memories of how I spent the morning having cornflakes for breakfast with my kids. Indeed, pattern separation and pattern completion are thought to be two fundamental and complementary functions of the hippocampus (Bakker et al. 2008).

Different variants of this basic idea have been presented in the literature, with different levels of biological realism. The more realistic versions of these models also make concrete hypotheses about how various neuromodulators (dopamine, serotonine, and

Decision Making 33

noradrenaline) contribute to both health and disease of memory functioning (Hasselmo, 1999; Rolls et al., 2008).

Besides the hippocampus, cortical areas have also been conceptualized via the language of Hopfield attractors. For example, so-called feature-based models of human categorization propose that each category that humans distinguish (say, cats) basically consists of a set of connected features (in the example, pet, hairy, claws, Facebook pictures). Such models are powerful tools because they show many desirable properties of human categories (McRae, 2004). In particular, presenting a sufficiently large subset of the features of a category can lead to pattern completion toward the full pattern of features of that category. In this way, the category (here, cat) can be identified in such models as simply corresponding to the attractor in feature space. Furthermore, presentation of a more typical exemplar will lead to faster and less error-prone completion of the attractor. This is again a useful property for a model of human categories because human categories also show this property: a prototypical class member (e.g., robin in the category of birds) will be recognized much more quickly than a less typical member (e.g., chicken in the category of birds) (McLeod et al., 1998). Finally, prototypical class members are more robust: Adding some random noise to the network weights (called a *network lesion*) leads first to a loss of less prototypical category members, and only later (i.e., with more lesions) to the loss of prototypical ones. This again corresponds to how human memory deteriorates in disease (e.g., dementia) (Rogers et al., 2004).

The attentive reader may consider at this point: this may all be nice and plausible, but how are these patterns encoded into the weights of the Hopfield model? How does one choose the weights w_{ij} in the first place? The topic of learning these weights is discussed in chapter 3.

The Diffusion Model

A very popular model that minimizes energy in activation space is the *diffusion model*. It was originally developed in physics in the 1920s to account for Brownian motion, the random motion made by a particle submerged in water. The model was popularized in psychology by Ratcliff and coworkers across the last few decades (Bogacz et al., 2010; Ratcliff, 1978). The basic idea behind the diffusion model is quite simple. Consider again the cat and dog detectors in the minimal model. In the diffusion model, one considers the *difference* between the activations for the cat and dog detector (e.g., $d = y_{\text{dog}} - y_{\text{cat}}$). Specifically, in the current notation, one can write the diffusion model as

$$d(t) = d(t-1) + v + N(t). \tag{2.8}$$

When the difference $d(t) = y_{\text{dog}}(t) - y_{\text{cat}}(t)$ reaches an upper *threshold*, the model decides that the stimulus is a dog; when it reaches a lower threshold, the stimulus is classified instead as a cat. As before, the variable $N(t)$ indicates noise from a normal distribution with mean 0. Figure 2.6 depicts the diffusion model. It illustrates the two thresholds: reaching first the upper threshold denoted a, or Right Response, would correspond to a dog response in this example; reaching first the lower threshold denoted –a, or Left Response, would correspond to a cat response. Figure 2.6 also illustrates two possible trajectories for the difference d, one arriving at the upper threshold, the other arriving at the lower threshold. On top of each of the two thresholds, an RT distribution for trajectories that reached that threshold is depicted (weighted with the probability of reaching each threshold); in the example, the Right Response is more probable and therefore has more probability mass.

A very important parameter in the diffusion model is *drift rate v*, also appearing in equation (2.8) and figure 2.6; this is the average slope of the decision trajectory. In the "cats and dogs" model, the drift rate corresponds to the difference in input for cat and dog detectors, so $v = in_{dog} - in_{cat}$. When a dog is presented, it is typically the case that $in_{dog} - in_{cat} > 0$, so drift rate v will be positive (v > 0). The trajectory is therefore more likely to hit the upper (dog) threshold first. When a cat is presented, drift rate will instead typically be negative (v < 0), and the trajectory will be more likely to hit the lower (cat) threshold first.

Another important parameter is the *starting point*. The starting point can be exactly halfway between the two thresholds [formally, in Figure 2.6, this would correspond to $d(0) = z = 0$], meaning that there is no bias toward either cats or dogs. If the starting point z is larger than 0, the energy minimization process starts closer to the dog (Right Response) threshold and is thus biased toward dogs. For completeness, I mention that also a *non-decision time* parameter is included in the basic diffusion model. This parameter captures all processes not related to the decision process. It is a constant that shifts the entire RT distribution rightward.

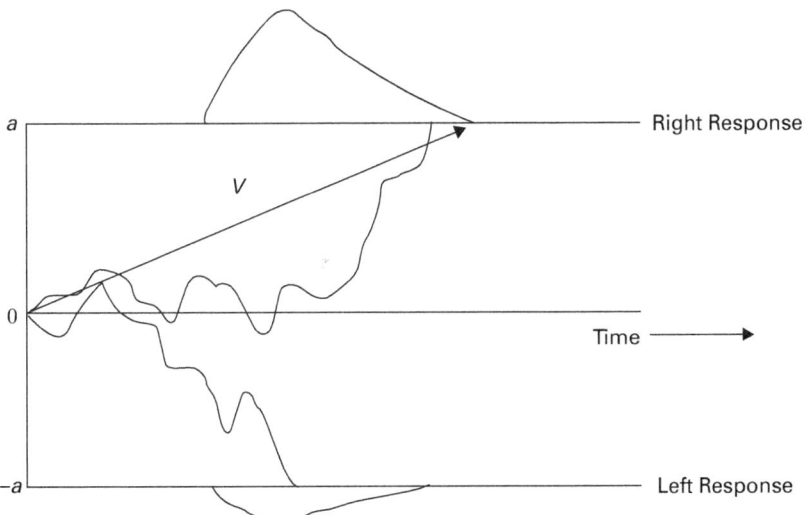

Figure 2.6
The diffusion model. This figure was reproduced and adapted with permission from White et al. (2011).

Just as for the other models described in chapter 2, the update rule of the diffusion model in equation (2.8) attempts to minimize an energy function defined across an activation space (Verdonck & Tuerlinckx, 2014). In reality, the diffusion model is slightly more complicated than I have presented it here because it operates in continuous time, not discrete time. I refer the reader to the specialized literature for delving into the details of the diffusion model (Ratcliff, 1978; Tuerlinckx et al., 2001).

The diffusion model has been very popular in cognitive neuroscience because it allows fitting choice (accuracy) and RT data simultaneously. In doing so, it disentangles a number of intuitively appealing concepts. Consider again the cat/dog categorization experiment. The stimulus quality or participant ability is expressed by the drift rate; easier stimuli (or better participants) have higher absolute drift rates ($|v|$). Catlike dogs and doglike cats would have drift rates close to zero; very prototypical cats and dogs would have large drift rates (in absolute value). The threshold parameter a embodies cautiousness, or a speed-accuracy trade-off. Higher thresholds entail slower but more accurate responses. Finally, the starting point embodies bias; a starting point z above 0 means a bias toward dogs. Indeed, when the starting point is above 0, the model is more likely to hit the upper threshold first. In contrast, a starting point below 0 would mean that the subject is biased toward cats. Thus, when an experiment has a sufficient number of trials and thus yields a sufficient amount of data, one can estimate the diffusion model parameters per subject, including each participant's ability (drift rate), cautiousness (threshold), bias (starting point), and non-decision time. In chapter 6, I will explain how this estimation can be carried out.

Besides the four basic parameters threshold, drift rate, starting point, and non-decision time, other parameters can be added to the diffusion model to better fit empirical data. For example, it might be reasonable to assume that parameters are not fixed but instead randomly vary from trial to trial. The amount of variability of such randomly varying parameters would then be another parameter (e.g., formalized as the standard deviation of a normal distribution). Such trial-to-trial variability is often included in the diffusion model for the drift rate, starting point, and non-decision time parameters (Ratcliff & Tuerlinckx, 2002). For the threshold parameter, such variability is not typically included. Instead, it is sometimes assumed that participants become less cautious (more urgent) as the trial goes on; this assumption can be captured by letting the upper and lower thresholds gradually shrink toward zero, such that reaching the threshold becomes easier (Cisek et al., 2009).

The Diffusion Model in Psychology

Using the diffusion model, Mulder et al. (2012) investigated how subjects implement bias in two-choice decision making. In this study, subjects decided on each trial whether a cloud of dots was either moving leftward or rightward. Before each trial, subjects were experimentally biased toward either a left or right response. In some experimental blocks, bias was induced on a trial-by-trial basis by making one response more likely (i.e., by informing the subject that it is more likely that the next correct response will be a leftward one); in other experimental blocks, it was induced by differential payoff to the two sides (i.e., by informing the subject that a correct response to the left will be given more reward than a correct response to the right). Both instructions would bias subjects toward responding with a left response.

However, how do subjects implement such a bias? As mentioned previously, subjects may implement bias by setting the starting point closer to the alternative toward which they are biased. Thus, they could have a starting point closer to the correct response in congruent trials (i.e., when the prior cue gives the correct indication of where the subject must move in the upcoming trial), and closer to the incorrect response in incongruent trials. Alternatively, subjects could have an increased drift rate in congruent trials and a decreased drift rate in incongruent trials. Mulder et al. (2012) found clear evidence for an effect in the starting point, not in the drift rate. This example shows how modeling can address questions that would be difficult to address otherwise, such as "How do subjects implement bias?" Furthermore, when the authors included subject-specific estimates of bias in their fMRI analysis, they observed that specific regions in a frontoparietal network were (across subjects) correlated with bias. Subjects with a stronger behavioral bias had higher activation in those areas. Thus, modeling can also help in finding the neural substrates of cognitive processes.

As another example, White et al. (2010) describe an application of the diffusion model to understand clinical disorders. They asked subjects, some low and others high in trait anxiety, to perform a simple, two-choice decision task. They fitted the diffusion model separately to postcorrect trials (i.e., trials where the previous trial was correct) and to posterror trials (i.e., trials where the previous trial was an error). As a result, they were able to investigate which parameters were changed after error trials and whether this process was different in participants with low versus high trait anxiety. Participants with high anxiety were found to increase their threshold after an error, whereas participants with low anxiety did not. Again, answering or even formulating such subtle questions would be hard without a computational model at one's disposal.

In summary, in this chapter, we have explored a number of models that were developed in the scientific literature with a variety of purposes. One common thread runs across those models: they all minimize an energy function via changing the activation (denoted x or y) of a collection of interconnected variables (or units). By doing so, these models change activation over time, and the resulting activation dynamics (e.g., the model choices or RTs) can be compared with empirical data. Exactly how model and empirical data can be statistically compared will be discussed in chapter 7. For now, there is a more pressing problem: All these models contained parameters (typically denoted as w_{ij}), and these parameters must be fixed in some way before the update rules laid out in this chapter can be set in action. A large part of this book, including the next chapter, will be devoted to the study of how these weights can be changed via optimization of their own goal function.

3 Hebbian Learning

History always matters.
—Esther Thelen

The Hebbian Learning Rule

In chapter 2, we constructed a basic pet detector with weights that were set by hand. In our toy example, that was still a reasonable assumption. We could use prior knowledge for setting the weights between "has its picture on Facebook" and the cat detector high. Obviously, this is not always going to work in real life. Besides, the reason we want a pet detector in the first place is to help making cat-versus-dog decisions. If we have to hardwire what is typical for cats and what is typical for dogs, the detector is not likely to help us much in making such decisions.

To change those weights, they will need to be learned. The current chapter discusses one type of learning—namely, *Hebbian learning*. One of the most celebrated learning rules, this is named after the Canadian neurophysiologist Donald Hebb. Note, first, that changing weights typically occurs at a slower time scale than the time scale of changing activation. When we divide time in discrete trials labeled $n = 1, 2, \ldots$, we can consider activation to change at a time scale *within* each trial. In contrast, changing the weights occurs at a time scale *between* trials. Figure 3.1 illustrates the difference between the two time scales. In real life, of course, there is no artificial division between time scales. However, the distinction helps one to focus on either activation change (fast time scale) or weight change (slow time scale).

I label input unit activation as x_j (with feature j indicating whether the animal has a picture on Facebook, has four legs, and so on); and output unit activation as y_i for output unit i. In this case, the correct output unit activations are not computed from the input units. They are instead labels provided by an external supervisor. One therefore calls this an instance of *supervised learning*. More generally, learning is supervised when the correct labels are provided by someone (a supervisor) outside the network. This chapter through chapter 5 discuss various supervised learning rules.

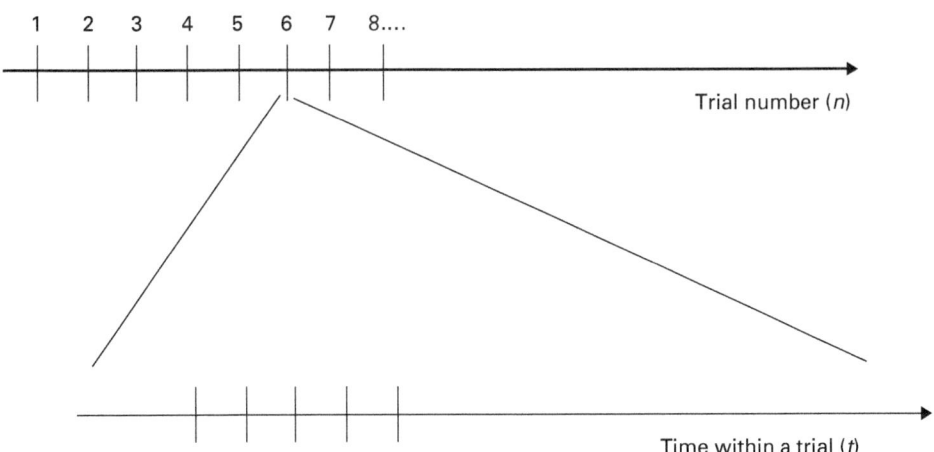

Figure 3.1
Fast time scale of activation (bottom) and slower time scale of learning (top).

As in chapter 2, I start from an energy function that needs to be minimized. This energy function will guide us in our search for a learning algorithm (i.e., for a recipe for changing the weights w_{ij}). Consider the following energy function:

$$E = -\frac{1}{I} \sum_{i=1}^{I} t_i y_i, \tag{3.1}$$

where the summation is taken across all I output units. The summation should in principle also be taken across trials, but I currently ignore that for simplicity (see also chapter 4). Here, y_i is the predicted value of target t_i (provided by the external supervisor), as predicted by the linear input, provided by J input features as follows:

$$y_i = w_{i1}x_1 + w_{i2}x_2 + .. + w_{iJ}x_J.$$

One can make E in equation (3.1) small by making y_i large whenever the required target output t_i is also large. As a simple example, suppose that there are just two output units, and the supervisor says that only the first output unit must be active (i.e., $t_1 = 1$, $t_2 = 0$). The energy function in equation (3.1) then becomes $-y_1/2$ (why?). If only one of the output units is allowed to be activated (so that either $y_1 = 1$ or $y_2 = 1$ but not both—I'll get back to this assumption later), the best (i.e., with E minimized) choice for y_1 and y_2 is clearly $y_1 = t_1 = 1$ and $y_2 = t_2 = 0$. Thus, in this case, we minimize the energy function if we make the required target output (provided by the supervisor) t_i and internally predicted output y_i similar. This is the rationale behind the energy function E in equation (3.1): we want to change the weights w_{ij} in such a way that the output from the model (y_i) is as similar as possible to the required output (t_i). To minimize this function, we again resort to the steepest descent rule.

To describe the weight change between two trials $n-1$ and n, I use the notation

$$\Delta w_{ij} = w_{ij}(n) - w_{ij}(n-1).$$

Hebbian Learning 39

If one applies the gradient descent principle from chapter 1 to the energy function [equation (3.1)], one ends up with the following Hebbian learning rule:

$$\Delta w_{ij} = \beta x_j t_i. \tag{3.2}$$

Note that I now am using β as the parameter with which to scale the step size (rather than α, as used in chapters 1 and 2). This emphasizes that we are updating weights rather than activation values. Because we are now learning (weights), I call β the *learning rate*.

If x_j, t_i are binary (0 or 1), then this learning rule says that the weight w_{ij} increases only if both the input unit (x_j) and required output (t_i) are active simultaneously (i.e., $x_j = t_i = 1$). Accordingly, Hebbian learning is sometimes characterized as "If it fires together, it wires together."

To see how this works, suppose that all weights start at zero. On trial $n = 1$, only the input feature "has its picture on Facebook" is active, and the supervisor informs the learner that the current animal is a cat. The Hebbian learning rule will then make sure that the connection between these two concepts ($w_{\text{cat},1}$) increases (because $x_j = t_{\text{cat}} = 1$). In contrast, the connection between "has its picture on Facebook" and the dog detector, for example, will remain at zero, because $t_{\text{dog}} = 0$. Suppose that then at trial $n = 2$, the same animal is presented, but no pet label t_i is presented. However, now we have learned something, and we can apply the linear principle to see what the predicted response is in the cat detector for the given input pattern:

$$y_{\text{cat}} = w_{\text{cat},1}x_1 + w_{\text{cat},2}x_2 + w_{\text{cat},3}x_3 \tag{3.3}$$

Because $w_{\text{cat},1} > 0$, it follows that $y_{\text{cat}} > 0$ (why?). Similarly, $y_{\text{dog}} = 0$ (why?). Thus, the Hebbian learning rule has helped solving the problem of knowing what the weights must be. Different input patterns will each be associated to either the cat or dog response.

Although learning provided the correct response in this rather small-scale example, this form of Hebbian learning is usually not such an efficient algorithm. The astute reader may have noticed this already by inspecting the energy function [equation (3.1)] that I started from. I claimed that this function can be minimized by choosing t_i and y_i to be as similar as possible, but to make this argument, I had to assume the rather restricted case where $y_1 = 1$ or $y_2 = 1$, but not both. Dropping this assumption, there are actually many (often inappropriate) ways to make this energy function very low. For example, one can let all weights go to infinity, and then E will keep on decreasing. And indeed, application of the rule in equation (3.2) will cause all weights to grow to infinity so long as each pair of concepts is presented together from time to time. In general, Hebbian learning only works well under very restricted conditions (see the next section). The reason why the Hebbian learning rule is nevertheless very popular (including in this book) is that it constitutes a building block of virtually all useful learning rules. This is because its general form can be written like this:

Change of weight between units i and j = function of unit i × function of unit j (3.4)

40   Chapter 3

For example, in the Hebbian learning rule [equation (3.2)], we can define

   function of unit $i = t_i$

and

   function of unit $j = \beta x_j$,

which makes it an instance of equation (3.4). But we can define other functions as well in equation (3.4), and thus end up with a large collection of learning rules. As a simple example, in the Hebbian learning rule as defined thus far, if activation is always 0 or 1, weights w_{ij} can never decrease. This is a problem because weights will quickly saturate in such a model; they will all be high, so there will be little differentiation among the various output patterns. All input patterns activate all output units. This can be remedied by subtracting a constant from each output unit; for example, in equation (3.2), we can replace t_i with $(t_i - 1/2)$, so the final update can be negative (in case $t_i = 0$), in which case weights can also decrease. This would also be called a Hebbian learning rule, and obviously it is also of the general form shown in equation (3.4). The novel term ½ can be considered a baseline to which the target is compared. In chapters 4 and 5, we will consider more adaptive baselines to which t_i is compared, where the baseline will consist of an internal prediction that the model makes about t_i. And more generally, as we will see in later chapters (4, 5, and 8–11), almost all learning rules can be written in a form similar to equation (3.4).

Exercise

3.1 Derive Hebbian learning [equation (3.2)] from its energy function [equation (3.1)].

When the information contained in "function of unit i" and "function of unit j" is available at synapse w_{ij}, we say that the learning rule is *local*. For example, in the Hebbian learning rule, the function of unit j is just βx_j, and the function of unit i is just the target t_i. Synapse w_{ij} contacts both x_j and t_i (it receives x_j internally and t_i from the supervisor), so the Hebbian learning rule is a local learning rule. Local learning rules are preferred because they are biologically plausible, in the sense that the weight has all the relevant information at hand to implement the rule.

Biology of the Hebbian Learning Rule

Direct evidence for the biological plausibility of Hebbian learning was found in the second half of the twentieth century by several research labs. Perhaps the most prominent is the work of Erik Kandel and colleagues, who received the Nobel Prize for Physiology or Medicine in 2000 for their study of the biological basis of learning. In a series of classical conditioning experiments, they used an invertebrate subject (the California sea hare, *Aplysia californica*), although the same principles have been studied in vertebrates as well. In this animal, Kandel and colleagues demonstrated that the connection between pre- and postsynaptic neurons grows during the conditioning experiment in a way predicted by Hebbian learning principles (Kandel, 2001). Specifically, it was already known that one type of receptor (the AMPA receptor, located at the postsynaptic cell) transmits messages from the presynaptic to the postsynaptic cell. For example,

Hebbian Learning 41

when glutamate (the main excitatory neurotransmitter) binds to the AMPA receptor, the AMPA receptor opens and sodium ions can flow into the postsynaptic cell. If a sufficient number of these AMPA receptors open, this will eventually cause an action potential to be generated at the postsynaptic cell.

But there is a second type of cell, the NMDA receptor, which behaves differently. The NMDA receptor is usually blocked by magnesium. However, if there is a sufficiently strong depolarization (i.e., signal at the postsynaptic cell), the NMDA receptor is unblocked. If glutamate binds to the NMDA receptor at that time, calcium can flow into the postsynaptic cell, via the NMDA receptor. The latter receptor thus functions as a coincidence detector (Caporale & Dan, 2008), and the Hebbian learning principle predicts exactly that: only if there were a presynaptic signal (glutamate) *and* a postsynaptic signal (magnesium unblocked) would there be a calcium effect. The consequence of calcium flowing in is that it influences synaptic strength at the postsynaptic level by increasing the number of AMPA receptors (Kauer & Malenka, 2007). As a result, information transmission will become more efficient. In modeling terminology, we say that weight w_{ij} has increased. One important difference between the Hebbian learning rule as discussed here and biological learning is that the latter is strongly timing-dependent, whereas the former is not. In particular, in biological systems, the postsynaptic strength will only increase if the presynaptic signal arrives before the postsynaptic signal (Caporale & Dan, 2008). The mechanistic reason for this is thought to be that a presynaptic signal before a postsynaptic signal causes a large calcium influx (as just discussed), whereas a postsynaptic before a presynaptic signal causes a small calcium influx, which triggers a different biological pathway. This makes computational sense, because a presynaptic signal before a postsynaptic one can mean that the presynaptic signal is causal for the postsynaptic one, but not if the order is reversed. This is different from the simple symmetric Hebbian learning principle discussed here, although time-dependent generalizations of the Hebbian learning rule have been proposed (Cooper & Bear, 2012).

Other research showed that manipulating the NMDA receptors also has behavioral consequences for the animal, as predicted by a Hebbian learning principle. For example, blocking the NMDA receptors impairs learning and stimulating the NMDA receptors actually improves learning (Tang et al., 1999). I cannot do full justice in this discussion to the deep biological discoveries by these research groups; suffice it to say that the general finding is that many of the discovered mechanisms are fundamentally Hebbian, and more complex learning builds on that process. I refer interested readers to Squire and Kandel (1999) for a nontechnical introduction to this literature. Of course, this does not imply that there are no non-Hebbian learning principles in the brain. However, it is theoretically satisfying that the (bottom-up) discovered biological rules and the (top-down) postulated computational rules show such a remarkable synergy.

Hebbian Learning in Matrix Notation

Next, I will generalize the notation and write the Hebbian learning rule in matrix format. The argument that follows in this section is based on Anderson et al. (1977). The matrix formalization serves two useful points. First, at a technical level, it provides some familiarity with the matrix notation of the linear principle. When you program your own model (e.g., in Python), it is handy to flexibly think in matrix terminology. Second, it also offers substantial insights, such as when various types of information will interfere with each other after storage in a network (and when not); and how network weights typically reflect the full history of an organism. Before I can explain this, some notational work needs to be done.

Following standard conventions, both vectors and matrices will be written in **bold**; vectors will typically be written in lowercase (e.g., $\mathbf{x}$) and matrices in uppercase (e.g., $\mathbf{W}$). Let us assume that we have a set of input units with activation x_j each, for J features altogether. We collect them into a vector $\mathbf{x}_n = (x_1, \ldots, x_J)$ on trial n, with the corresponding classification t_n (i.e., Cat or Dog). To go from an input pattern $\mathbf{x}_n$ to a classification of Cat versus Dog, we employ the linear principle and weights w_{ij}. Weights are collected in weight matrix $\mathbf{W}$. According to the linear principle, we can write the transformation from $\mathbf{x}_n$ to the input for the two pet detectors $\mathbf{in}_n = (\text{in}_{\text{Cat}}, \text{in}_{\text{Dog}})$ as $\mathbf{in}_n = \mathbf{W}\mathbf{x}_n$ (see exercise 3.3).

Next, consider learning. This is supervised learning, and thus our supervisor provides the correct labels (targets) $\mathbf{t}_k$ for input patterns $\mathbf{x}_k$. Just as we can write the linear principle in matrix format, we can also write the learning rule [equation (3.2)] in such a format (see exercise 3.4), where $\mathbf{t}_n = (t_1, \ldots, t_J)$ is the target vector on trial n. In particular,

$$\Delta \mathbf{W} = \mathbf{W}_n - \mathbf{W}_{n-1} = \beta \mathbf{t}_n \mathbf{x}_n^T. \tag{3.5}$$

Now, ignore for the moment the learning rate β; or if you find that problematic, assume that $\beta = 1$. Then repeated application of equation (3.5) (in particular, n times because we have n trials) leads to

$$\mathbf{W}\mathbf{x} = \sum_{n=1}^{N} \mathbf{t}_n \left(\mathbf{x}_n^T \mathbf{x} \right). \tag{3.6}$$

Try to convince yourself that the updated equation (3.5) leads to the more explicit formulation in equation (3.6) (on the condition that the initial weight matrix consists of zeroes).

In its simplicity, this compact formula reveals a great deal about neural networks trained by Hebbian learning. First, it shows that dissimilar input patterns will not interfere with each other (with "dissimilar" to be defined in a moment). Let's say, for example, that we have orthonormal vectors $\mathbf{x}_1$ and $\mathbf{x}_2$. This means that $\mathbf{x}_1$ and $\mathbf{x}_2$ are orthogonal (i.e., $\mathbf{x}_1^T \mathbf{x}_2 = 0$); and $\mathbf{x}_1$ and $\mathbf{x}_2$ are normalized—that is, each vector has length 1 (i.e., $\mathbf{x}_k^T \mathbf{x}_k = 1$, for $k = 1, 2$). See figure 3.2 for an illustration of two orthonormal vectors. Suppose further that only those two vectors are trained using Hebbian learning. If we present vector $\mathbf{x}_1$ after training, the response of the network will be [applying equation (3.6)]:

$$\mathbf{W}\mathbf{x}_1 = \mathbf{t}_1. \tag{3.7}$$

So, the input to the cat and dog detector units is just the earlier label $\mathbf{t}_1$. Similarly, if we present $\mathbf{x}_2$ to the network, it will respond as

$$\mathbf{W}\mathbf{x}_2 = \mathbf{t}_2.$$

Hebbian Learning   **43**

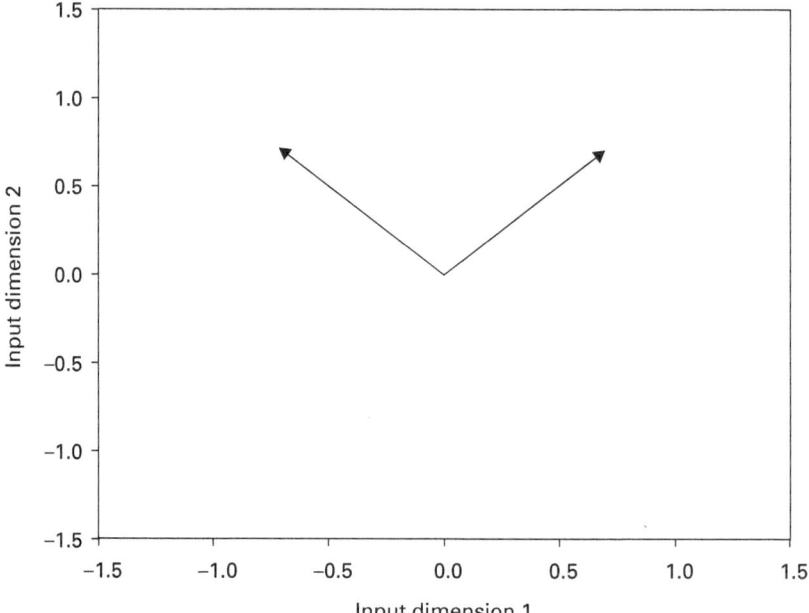

Figure 3.2
Two orthonormal (orthogonal and length 1) input vectors.

In other words, the network generates the correct output to both x_1 and x_2 as input. It has learned the information! In general, though, the prospects for the model trained by a Hebbian algorithm are not so rosy. The correct labels t are produced in this case because vectors x_1 and x_2 were chosen to be orthonormal. Normalization is not a strong assumption; any set of vectors can easily be normalized. But orthogonality is a very strong assumption: Vectors are typically not orthogonal. In fact, if we sample two vectors at random, the probability that they turn out to be orthogonal is close to zero. This is, of course, also true for normality (in the sense just defined, as being of unity length), but it can be remedied for each vector separately (just divide each element of each vector x_k by $(x_k^T x_k)^{1/2}$). Orthogonality, instead, requires checking interactions between vectors. This is why it is a strong assumption.

For completeness, let me add that this is mainly true in low-dimensional spaces: Orthogonality is less an issue in high-dimensional spaces, and in general, Hebbian learning works better in higher dimensions (see also the discussion of Hopfield model later in this chapter). But for a fixed number of dimensions, it is less efficient than the learning rule that will be discussed in chapter 4.

Also, in K dimensions, at most K vectors can be mutually orthogonal, so that is a strict upper limit on how many patterns can be stored using Hebbian learning in this case. In practice, capacity [in the sense of storing vector pairs (x, t) in memory] deteriorates quickly using Hebbian learning as more patterns are stored. To illustrate this point, in the next section, I will work out network capacity in a slightly different setting of the Hopfield model than what we already encountered in chapter 2.

I can now return to my promise at the start of this paragraph—namely, that this notational system helps us to see how weights reflect the full learning history of an organism. Note, first of all, that vectors will interfere with one another to the extent that they are similar. We can view $x_n^T x$ in equation (3.6) as a similarity function between vectors x_n and x. Indeed, note that if x_n and x have mean zero and $x_n^T x_n = x^T x = 1$, then $x_n^T x$ is simply the Pearson correlation between x_n and x. We can accordingly rewrite equation (3.6) as

$$Wx = \sum_{n=1}^{N} t_n \text{sim}(x_n, x).$$

Thus, when a new vector x is presented, the model will evaluate the novel object embodied in x and compare it to all the stored x_n values via the similarity function $\text{sim}(x_n, x)$. These similarity values are then multiplied by the desired output pattern on trial n (namely, t_n). The final result is a weighted sum of t-vectors, weighted by the similarity of novel vector x to the stored vectors x_i. Informally, every response that a network provides is a function of everything it has ever experienced in the past. Even more informally, history always matters. Exercises 3.2–3.6 are intended to give you some hands-on experience with Hebbian learning in matrix notation.

Exercises

3.2 If necessary, refresh your knowledge of linear algebra, by consulting a suitable source such as xaktly.com.

3.3 Show that the linear activation rule [equation (3.3)] can be written as **in = Wx**.

3.4 Show that the Hebbian learning rule [equation (3.2)] can be written as in equation (3.5).

3.5 Show that equation (3.7) follows from equation (3.6) if vectors x_1 and x_2 are orthonormal (i.e., orthogonal and normalized).

3.6 Write a program to convince the skeptical researcher that orthonormal input vectors can always be learned; but that classification quickly goes wrong when input vectors are not orthonormal.

Memory Storage in the Hopfield Model

Until now, I have considered a supervised-Hebbian two-layer model, with activations x and y at the input and output layers, respectively. However, the concept of Hebbian learning can be applied in other models (with other energy functions, activation functions, and architectures) too. Now, let's use our hard-won knowledge of the matrix notation of Hebbian learning to understand memory storage and when it can break down in the context of the Hopfield model discussed in chapter 2. In particular, how many patterns can be stored in a Hopfield model before the patterns start interfering with each other? I will follow the argument of Rojas (1996), who considers the slightly simpler case in which all threshold parameters $\theta_i = 0$; however, this causes no loss of generality, as exercise 3.7 shows. Moreover, all activation values x are -1 or 1 rather than 0 or 1. The derivation is similar for the 0/1 case if we add the assumption that each vector has the same length.

Hebbian Learning

When pattern $\mathbf{x}$ is presented, the Hebbian learning rule when applied to the Hopfield model, simply consists of the rule

$$\Delta w_{ij} = \beta x_j x_i$$

for the connection from unit j to unit i. However, we do not want a unit to connect to itself, so we restrict the learning rule to the case where $i \neq j$. If we apply this rule repeatedly to the patterns {John, male, poor} and {Mary, female, rich} from chapter 2, these patterns will become attractors of the Hopfield model. For example, connection $w_{\text{male,John}}$ will increase by an amount β, which is the learning rate, if the first pattern is presented.

How many patterns can be stored in this way? Consider the weight matrix after a single pattern $\mathbf{x}_1$ has been presented:

$$\mathbf{W}_1 = \mathbf{x}_1 \mathbf{x}_1^T - \mathbf{I}. \tag{3.8}$$

Here, the learning rate β is again ignored for simplicity (or assumed that $\beta = 1$). In equation (3.8), $\mathbf{I}$ is the identity matrix with $\mathbf{I}_{ii} = 1$ and $\mathbf{I}_{ij} = 0$ if $i \neq j$. Subtracting this matrix in equation (3.8) is performed to implement the requirement that a unit will not become connected to itself.

Let's now consider which activation patterns $\mathbf{x}$ the model will be attracted to. The energy function that we will consider for pattern $\mathbf{x}$ is exactly the same as the one we already introduced for the Hopfield network in chapter 2 [i.e., equation (2.6)]. When only this single pattern $\mathbf{x}_1$ is learned, we can write the resulting energy function for any novel pattern $\mathbf{x}$ in the following way:

$$E(\mathbf{x}) = -\left(\mathbf{x}^T \mathbf{x}_1\right)^2 / 2 + J/2, \tag{3.9}$$

which reaches a minimum when $\mathbf{x} = \mathbf{x}_1$. Hence, $\mathbf{x}_1$ is a stable state, and novel patterns $\mathbf{x}$ will be attracted toward $\mathbf{x}_1$. Here, *any* input pattern will eventually be attracted toward $\mathbf{x}_1$. This is not particularly useful; see figure 3.3 for an illustration of what would happen in this model. The model has stored pattern $\mathbf{x}_1$, and any environmental input as presented via a novel input pattern $\mathbf{x}$ will bring $\mathbf{x}_1$ to the model's mind.

So next, let's try to store an additional $N - 1$ patterns (i.e., N patterns altogether). Applying the learning rule from equation (3.5) repeatedly (under the same assumptions as before), the weight matrix becomes

$$\mathbf{W} = \mathbf{x}_1 \mathbf{x}_1^T + \mathbf{x}_2 \mathbf{x}_2^T + \cdots + \mathbf{x}_N \mathbf{x}_N^T - N\mathbf{I}.$$

In this case, we can hope to have a situation as in figure 3.4: If we present a perturbed version of some pattern presented in the past, it will remind the model of the corresponding pattern. The model has thus stored the appropriate memory. The fact that this will happen can be seen by considering what happens if we present the old pattern $\mathbf{x}_1$ again to the model:

$$\mathbf{W}\mathbf{x}_1 = (J - N)\mathbf{x}_1 + \sum_{j>1} \text{sim}(\mathbf{x}_j, \mathbf{x}_1)\mathbf{x}_j. \tag{3.10}$$

Recall that each unit has a threshold of 0; if activation is smaller than 0, we set $x = -1$; if activation is larger than 0, we set $x = 1$. Therefore, if the number of stored patterns (N)

$X_1 =$

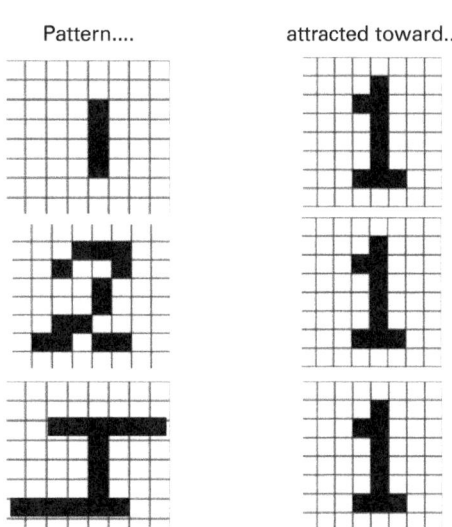

Pattern....   attracted toward...

Figure 3.3
Any pattern is attracted toward $\mathbf{x}_1$. This figure was reproduced and adapted with permission from www.codeproject.com.

is small relative to the number of units (so that $J - N \gg 0$) and the overlap between patterns is small (so that the terms of the summation in (3.10) are small), the first term on the right side of equation (3.10) will dominate, the elements of $\mathbf{Wx}_1$ will have the same sign as the elements of $\mathbf{x}_1$, and after thresholding, the activation pattern after one Hopfield update will become close to $\mathbf{x}_1$.

In general, a larger number of units and/or little overlap between various patterns $\mathbf{x}$ will be beneficial for memory storage. In practice, though, overlap between patterns will severely disrupt memory storage, as illustrated in figure 3.5. Using ch3_tf_hopfield.py (on the GitHub repository), I trained a Hopfield network on the well-known MNIST data set, a digitized collection of handwritten digits (each 28×28 pixels in size) that is often used for training and testing image classification models. I first trained the Hopfield model on just two of the images. After training, I started from a random input pattern (see row 1), and let the model evolve its activation pattern for three steps. Here, it is clear that one of the patterns is retrieved (a written "7"). Next, I trained the model on five of the written digits (row 2). In this case, the output pattern is an unreadable mixture between the various digits.

Exercises

3.7 In the Hopfield model, argue why the thresholds θ can be considered as being just another set of weights and thus be incorporated into weight matrix $\mathbf{W}$.

3.8 Derive the Hebbian learning rule as applied to the Hopfield model, from the Hopfield energy function shown in chapter 2 (where the function is considered to have $\mathbf{W}$ rather than $\mathbf{x}$ as its variables).

Hebbian Learning   **47**

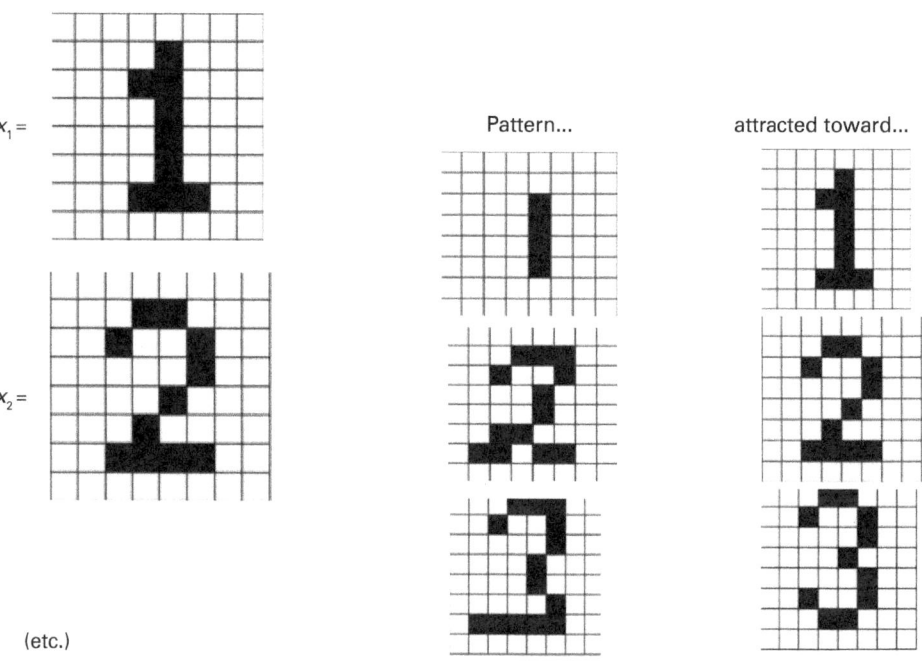

Figure 3.4
Small perturbations of $\mathbf{x}_i$ are attracted toward $\mathbf{x}_i$. This figure was reproduced and adapted with permission from www.codeproject.com.

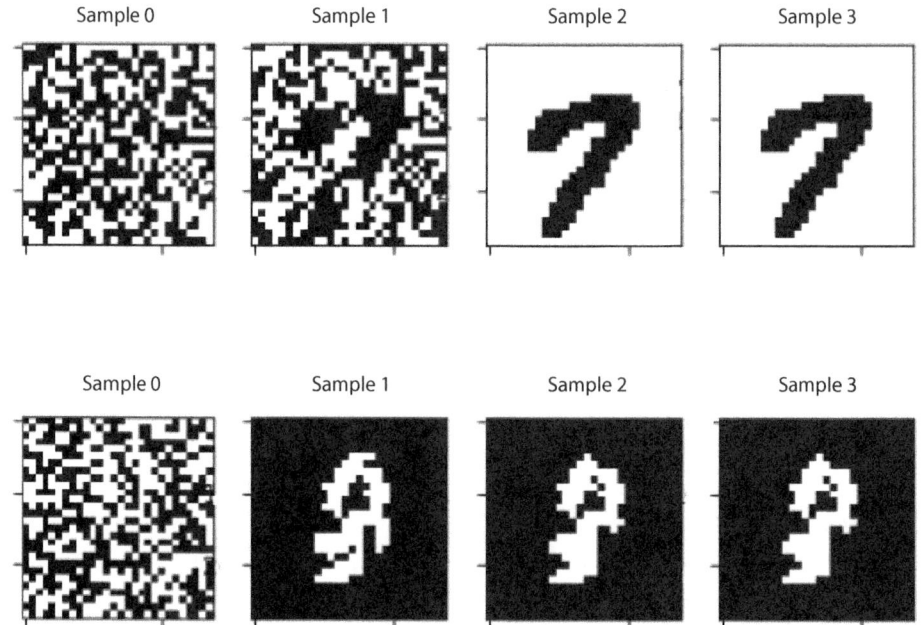

Figure 3.5
Row 1: After training on 2 items, starting in a random activation pattern leads to one of the stored digits (7). Row 2: After training on 5 items, starting in a random activation pattern, leads to an unreadable digit.

3.9 Derive equation (3.9) from the Hopfield energy function reported in chapter 2. Hint: First write the Hopfield energy function in matrix format.

3.10 Write both the Hopfield and supervised-learning Hebbian energy functions in matrix format. Do you see their formal similarity?

3.11 In figure 3.5, there is black-white figure-background inversion between rows 1 and 2 (i.e., black is figure in row 1, but background in row 2). Can you explain why?

Hebbian Learning in Models of Human Memory

The intuitive appeal, biological plausibility, and convenient linear structure of Hebbian learning have made it a common theme in models of human memory at both short-term and long-term scales. One example involves the feature-based Hopfield models of human concepts and categorization mentioned in chapter 2 (McRae, 2004). In this approach, each concept is characterized by a vector of features derived from feature norms where participants indicate whether each particular feature characterizes each particular concept (e.g., green, round, . . . for *apple*). Then, a Hebbian learning rule identifies appropriate weights between the units that represent the features. The resulting model can then be compared with a rich database that reports which concepts are easy to remember and respond to, how brain damage affects concept memory, and so on. For example, McRae et al. (1997) trained a Hopfield model with a Hebbian learning rule based on feature co-occurrences of several real-world objects. They observed that the model could account for semantic priming and similarly judgments between concepts. Incidentally, it is not the case that a model equipped with Hebbian learning rule would pick up only on first-order relations (correlations) between features. For example, consider the features "green" and "red" in the category of apples. Presumably, these features will be not or negatively correlated: An apple that is red will tend not to be green at the same time. Yet, because both features are correlated with other typical apple features (round, sweet), they can both become part of the apple attractor.

A crucial concept in memory research is context (Burgess & Hitch, 1999, 2005; Howard & Kahana, 2002). "Context" refers to all the features that cooccur with (but are not themselves part of) a specific memory. For example, a dinner can occur in the context of a specific restaurant or a specific living room. But context can also be more abstract: if a number of items is presented to be memorized, each item can be considered to be presented in a slightly different context. Such time-varying contexts have been considered crucial for explaining memory and its characteristics.

In the short-term memory[1] domain specifically, an early associative-chain model proposed that items are associated to each other (as in a chain: A → B → C → ..). However, the concept of an associative chain predicts that if, say, item B is not recalled, also the subsequent items are not. This prediction is not consistent with empirical data (Burgess & Hitch, 1999), and a context representation elegantly remedies this problem. Indeed, in context-based models, recall is mainly driven by the context, rather than by the previous item in the list. In context-based computational short-term memory models, context is typically implemented via a pointer, specifically a vector of activations ($\mathbf{x}$) to which an item in short-term memory can be bound. For example, suppose that we have to remember a list of items for our grocery shopping: apples, wine, pasta, carrots, sugar, and coffee. For remembering an ordered list like this, computational models of short-term memory assume that subsequent items are bound to subsequent pointers: $\mathbf{x}_{pointer1}$ is bound to $\mathbf{x}_{apples}$, $\mathbf{x}_{pointer2}$ is bound to $\mathbf{x}_{wine}$, and so on. In the simplest case, we can just consider each pointer vector to have 1 at its corresponding place: $\mathbf{x}_{pointer1} = (1, 0, \ldots, 0)$, $\mathbf{x}_{pointer2} = (0, 1, 0, \ldots, 0)$, and so on. But other types of pointer vectors are possible. For example, each pointer vector may consist of a superposition of oscillators at different frequencies (Brown et al. 2000). In yet other models, associations go in both directions (i.e., also from the items to the pointers; e.g., Howard & Kahana, 2002); the latter can thus be considered a hybrid between an associative-chain and pure context models. Finally, some other models code only the start and end locations as pointers, and associate all items in a list with differential strengths to these two pointers in short-term memory (Henson, 1998).

The pointer-item associations (weights) can be learned via Hebbian principles during the learning phase. For example, all features of pointer1 (collected in vector $\mathbf{x}_{pointer1}$) are presented simultaneously with all features of apples (collected in vector $\mathbf{x}_{apples}$); and this similarly applies for the other pairs. See figure 3.5 for a simplified illustration of this principle when the third item (pasta) is presented. As a result of Hebbian learning, cooccurring features will become connected. When the list needs to be reinstated (e.g., when one is in the store, trying to remember the items needed), it suffices to replay the full list of pointers ($\mathbf{x}_{pointer1}$, $\mathbf{x}_{pointer2}$, $\ldots$), such that at each pointer, its corresponding associated item can be recalled.

Application of this principle leads to interesting hypotheses about a number of robust behavioral properties in the study of short-term memory. For example, a very robust property is the frequency effect (i.e., items that are presented more often are remembered better). The frequency effect is obviously consistent with Hebbian learning. Connections that undergo more Hebbian learning, will tend to be stronger, and thus the corresponding item can be recalled better.

A subtler phenomenon is the primacy effect. This means that items near the start of the sequence are remembered better than later items; for example, apples would be remembered better than carrots. A related effect is the recency effect. As the name suggests, items that are presented the most recently (i.e., those near the end of the sequence) are remembered better. Together, primacy and recency lead to the typical bow shape of memory, where extreme (start and end) items are remembered better than items in the middle.

The pure Hebbian learning rule shown in equation (3.2) has no order sensitivity, and thus also no primacy or recency effect. However, both primacy and recency follow naturally from several models that combine Hebbian learning with context-to-item mappings (Hasselmo & Eichenbaum, 2005; Howard & Kahana, 2002; Page & Norris, 1998[2]). Consider the item in figure 3.5: Because a number of pointer units are activated,

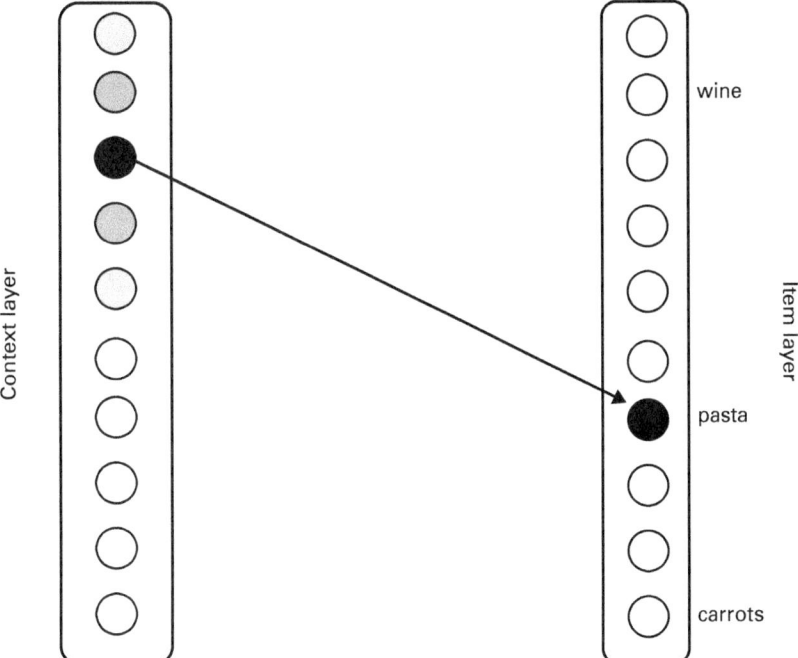

Figure 3.6
Illustration of the context model of memory (only a few items are labeled). The activation at the context layer is $x_{pointer3}$; pasta is the third item on the list given in this example. Unit shading indicates activation (with more black indicating more active). The arrow shows the weight that will be most updated after this pair of patterns is presented.

a few items in the item layer will typically be activated, causing interference between the items. The start and end items suffer less interference from their neighbors (they have just one neighbor, whereas all other items have two), leading to improved recall relative to items near the middle of the list. Although other mechanisms may contribute to primacy and recency (Davelaar et al., 2005), the concept of interference due to noisy context-item mappings appears in several human short-term memory models.

One such model is the "phonological loop" model of Burgess and Hitch (1999). The phonological loop is a memory buffer (Baddeley & Hitch, 1974) used for remembering phonological information on a seconds-to-minutes time scale (as when remembering groceries for shopping). The authors demonstrate how a Hebbian learning algorithm can be implemented in such a model that solves short-term memory tasks. When the subsequent items are learned (apples, wine, and so on), the model can produce them back afterward in the correct order.

An interesting aspect of this model is that it parallelizes serial order processing; in particular, when a number of items has to be recalled, all items compete in parallel, and the winner (the strongest item) is produced. This item then leaves the competition, at which point the remaining items compete, the winner is again produced, and so on. This process is called *competitive cueing*; recent data provide neurophysiological support

Hebbian Learning

for the existence of competitive cueing in the primate cortex (Averbeck et al., 2002; Kornysheva et al., 2019). Context-item mappings naturally lead to a process where items are put "in the queue" according to activation strength. Indeed, if the context gradually shifts, activation of the corresponding items shifts along with it. How competition between items can be implemented will be discussed in chapter 10.

One remarkable aspect of memory effects (such as frequency, primacy, and recency) is that they occur at several time scales. For example, they occur in short-term memory (a fast time scale), but also in long-term memory. An example is the bow-shaped curve (i.e., primacy and recency combined) one sees in the US presidents that subjects can recall by heart (Roediger & Crowder, 1976). For example, George Washington and Joe Biden are likely to be better known than Grover Cleveland or James Garfield. The fact that these effects (such as frequency, primacy, and recency) occur across time scales from seconds to years suggest that no simple (single-time-scale) architectural feature (such as a short-term memory buffer) is responsible for them. Instead, they suggest that basic computational principles apply across different time scales. It thus makes sense to implement short- and long-term associations via similar mechanisms.

This is exactly what happens in the Burgess and Hitch (1999) model. Here, both short- and long-term memories are formed via Hebbian learning, but short-term weights develop and decay much more quickly than long-term weights. Thus, both empirical associations and dissociations between short- and long-term memory phenomena can be accounted for. Of course, saying "There are both short- and long-term memory" is a bit too easy. Only by specifying and testing detailed computational models of short- and long-term memory and their interactions will we be able to determine to what extent short- and long-term memory are based on same or different mechanisms.

In summary, I have discussed one of the simplest rules for changing weight parameters w_{ij} in a neural network. This is Hebbian learning, often summarized by the quote "If it fires together, it wires together." I have shown how this rule can be biologically implemented, and how it can explain several basic characteristics of human memory. Although Hebbian learning in its most basic form is computationally inefficient (see the discussion we have had about figure 3.5), it serves as a building block for more sophisticated learning rules, including the error-based learning rules that are discussed next, in chapters 4 and 5.

4 The Delta Rule

To err is human.
—Seneca

The Delta Rule in Two-Layer Networks

In chapter 3, we used the Hebbian learning rule to train our pet detector. But it functioned well only under quite specific circumstances. In the next two chapters, we consider a number of learning rules that perform much better computationally, in the sense that they can learn much more about the input data they receive from the environment. I refer the reader who wants to know more about the technical background of chapters 4 and 5 to Bishop (1995).

In this chapter, we consider the delta rule. Just like the Hebbian learning rule, this rule can be used for a model architecture with two layers of units. Consider again figure 2.1a in chapter 2 or look at the more generic two-layer model shown in figure 4.1. Like before, we apply the linearity principle to calculate activation, arriving at the pet (or more generally, output) level:

$$in_i = \sum_{j=1}^{J} x_j w_{ij}. \tag{4.1}$$

Subsequently, we transform the incoming activation to make a prediction about which pet is presented (i.e., variable y):

$$y_i = f(in_i),$$

according to a function $f(.)$, called an *activation function*. The activation can minimally just be the linear input arriving at the pet nodes, in which case we had $y_i = f(in_i) = in_i$. In chapter 2, the activation y_i was defined implicitly by the energy minimization rule. In the Hopfield model presented in chapter 3, we also considered an activation function (without naming it as such): if a unit input was larger than its threshold, its activation was set to 1, otherwise to 0. In this chapter, we make the activation function explicit.

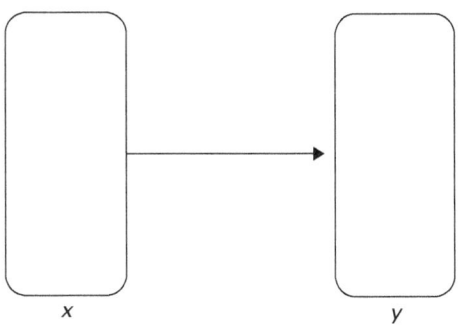

Figure 4.1
A generic two-layer model.

One popular activation function is the logistic function, defined as

$$y_i = \frac{1}{1 + \exp(-\gamma \times (in_i - \theta))}. \tag{4.2}$$

Here, θ is a threshold value; if in_i exceeds this threshold (i.e., $in_i - \theta > 0$), the value of y_i becomes larger than 0.5. The parameter γ is a slope parameter; the larger it is, the steeper the logistic function. We can consider the logistic activation function to be the soft version of the threshold function: If γ tends to infinity (i.e., the function becomes "harder"), we end up with a threshold activation function. In this case, if in_i is below the threshold, $y_i = 0$; if in_i is above the threshold, $y_i = 1$. Several activation functions are illustrated in figure 4.2.

In this model, we have parameters w_{ij}, θ, and γ, but some remarks are in order before we attempt to optimize them. First, parameter θ can be considered as bias parameter $w_{i,\text{bias}}$, attached to an input unit with constant activation $x_{\text{bias}} = -1$. This is very similar to how the thresholds in the Hopfield model discussed in chapters 2 and 3 can be treated; see exercise 3.7 in chapter 3. Indeed, in equation (4.1), where J input features are linearly combined, we can add a $(J+1)$th feature:

$$in_i = \sum_{j=1}^{J} x_j w_{ij} + (-1) \cdot \theta = \sum_{j=1}^{J+1} x_j w_{ij},$$

where $x_{J+1} = -1$ and $w_{i(J+1)} = w_{i,\text{bias}}$.

Second, parameter γ cannot be optimized at the same time as the w parameters because the optimization problem would be ill defined. Indeed, for any solution $(\gamma, \mathbf{w})$, the alternative solution $(\gamma/C, \mathbf{w}C)$, with C an arbitrary constant, is just as good a solution (see exercise 4.2). Therefore, γ is typically fixed prior to optimization. Nevertheless, γ also has some interesting psychological interpretations, as we consider in later chapters. Currently, we set it to $\gamma = 1$, and we therefore end up with the following logistic activation function:

$$y_i = \frac{1}{1 + \exp\left(-\sum_j x_j w_{ij}\right)}. \tag{4.3}$$

The Delta Rule   55

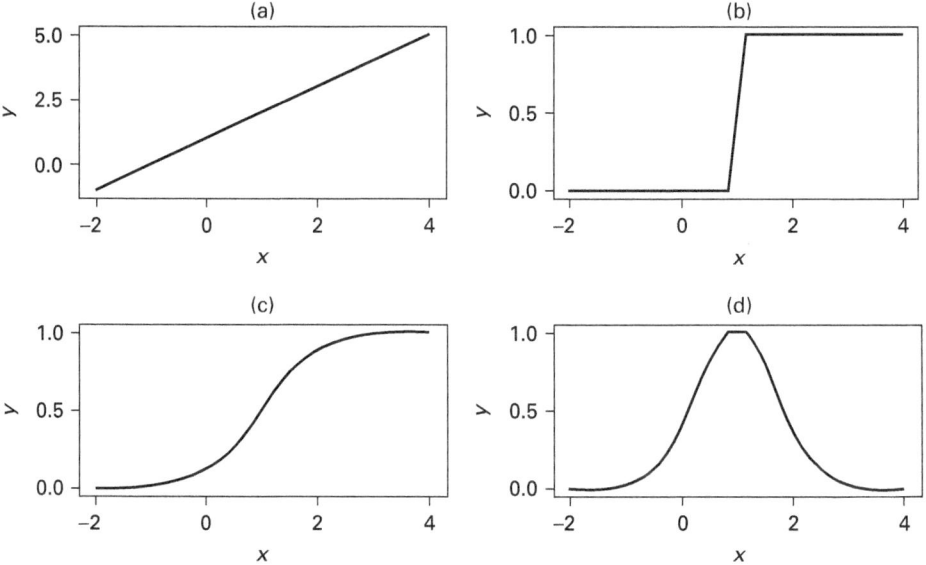

Figure 4.2
Plots of activation functions: (a) linear; (b) hard threshold; (c) soft threshold (or sigmoid); (d) Gaussian.

This is the activation function that I will use in optimization in the next section, as a function of weights w_{ij}.

Mean Square Error Function
The optimization function that was considered in chapter 3 had some clear deficiencies. I now consider the following mean square error (MSE) function:

$$E = \frac{1}{I} \sum_{i=1}^{I} (t_i - y_i)^2, \tag{4.4}$$

where again, t_i are the values provided by the external supervisor.

The summation in equation (4.4) should in principle also be taken across all trials (n). But on any trial, the model sees just one stimulus-category pair, so the error E that we actually optimize is restricted to just one such pair on each trial. For this reason, the current learning rule is sometimes also called *stochastic gradient descent* (where one stochastically samples just one stimulus-category pair on each trial). Thus, by the same reasoning, I also already applied in chapters 2 and 3 the stochastic version of gradient descent (or ascent).

Recall that in Hebbian learning (as discussed in chapter 3), we could minimize the error function by just making the values of y_i as large as possible. Note that this would not work in this case: here, the minimization goal can only be attained by making the predicted values (y_i) as similar as possible to the required (supervisor) values (t_i). We can therefore also expect to end up with a more powerful update rule (and we will, as

I will show very shortly). We square the error function in equation (4.4) because the direction of the error doesn't matter: a deviation of $t - y = 1$ is just as good (or as bad) as a deviation of $t - y = -1$. Squaring makes sure that either type of error contributes 1 to overall error.

I again apply gradient descent in weight space thus:

$$\Delta w_{ij} = -\beta \frac{\partial E}{\partial w_{ij}}. \tag{4.5}$$

From now on, I will use the notation $\partial f(x,y)/\partial x$, which is the partial derivative of function $f(x,y)$ toward variable x. It is a straightforward generalization from the single-variable $df(x)/dx$ toward functions of more than one variable.

It can be shown that working out equation (4.5) algebraically leads to

$$\Delta w_{ij} = \beta x_j (t_i - y_i) \frac{d}{d\text{in}_i} f(\text{in}_i). \tag{4.6}$$

In the case of the linear activation function, this becomes simply

$$\Delta w_{ij} = \beta x_j (t_i - y_i).$$

If instead we use the logistic activation function [equation (4.3)], the update rule becomes

$$\Delta w_{ij} = \beta x_j (t_i - y_i) y_i (1 - y_i) \tag{4.7}$$

To see why this is true, one can either look at Bishop (1995) or perform exercise 4.3. Doing that exercise has the advantage that you will understand and remember the concept better than if you just read about it.

Cross-Entropy Function

In several applications, it is natural to consider the output units as hypotheses. In the now-familiar pet scenario that we have discussed in this book, output units might represent the hypotheses "I'm looking at a cat" or "I'm looking at a dog." First, let's consider the case where each output unit represents a separate hypothesis; as a consequence, it is possible that an object is both a cat and a dog. This is rather odd, of course, but for simplicity, let's go with it anyway. We may then interpret the value of y_i as the predicted *probability* of the hypothesis (e.g., it's a cat, yes or no). The target values $t_i \in \{0, 1\}$ then state whether the hypothesis is true ($t = 1$; "I'm looking at a cat") or false ($t = 0$; "I'm not looking at a cat"). If we denote by y_i the probability that the target value of $t_i = 1$, then the probability of the data for unit i (t_i) can be written as

$$L = y_i^{t_i} (1 - y_i)^{1-t_i}.$$

The reason why I call this function L will become clear in chapter 6. For now, if we take the logarithm of this function (the reason of which will also be explained in chapter 6), and sum across all output units, we obtain the following result:

$$\log(L) = \sum_i t_i \log(y_i) + (1 - t_i) \log(1 - y_i). \tag{4.8}$$

The Delta Rule   57

We now attempt to maximize equation (4.8) as a function of the weights. If we again apply the gradient ascent principle, maximizing the function [equation (4.8)] leads to the following update rule for weights w_{ij}:

$$\Delta w_{ij} = \beta x_j (t_i - y_i).$$

This is called the *cross-entropy maximization rule* because equation (4.8) contains the cross-entropy (which is similar to a correlation) between vectors $(t_i, 1 - t_i)$ and $[\log(y_i), \log(1 - y_i)]$, for each value of i (Hinton, 1989). At a more practical level, an advantage of the current rule relative to the MSE rule [equation (4.7)] is that the cross-entropy maximization rule converges on the optimal point more quickly than the MSE minimization rule because the factors y_i and $1 - y_i$ no longer appear in the weight update equation. Indeed, due to the y_i and $1 - y_i$ factors in the MSE weight update equation (4.7), once a weight is shut off, it is very hard to reactivate it again because its gradient will be close to zero. For this reason, the cross-entropy rule may perform considerably better than the MSE rule in capturing complicated input-output mappings, such as those that occur, for example, in psycholinguistics (e.g., orthography-phonology mappings; see Plaut et al., 1996).

With the cross-entropy maximization rule, we consider the neural network model as a statistical model that finds the parameters (w_{ij}) that maximize the (log) likelihood (see chapter 6) of hypotheses (Is it a cat? Is it a dog?), given the data (i.e., the pattern of input activation). We will consider more systematic links between neural networks and statistical models in chapter 5, on multilayer networks; and in chapter 10, on unsupervised learning (specifically, Boltzmann machines in the latter).

In general, one can come up with other activation rules, activation functions, and error functions and subsequently apply gradient descent on the resulting model. Different combinations may (indeed will) have different properties, as I illustrated by comparing MSE and cross-entropy error functions. There is one restriction, however: because we apply gradient descent, the resulting energy (e.g., error) function must be differentiable. Thus, for example, the gradient descent algorithm is not applicable to a network with a threshold activation function.

Exercises

4.1* Derive an explicit form of the activation function $f(.)$ defined in chapter 2 (e.g., equations (2.4) and (2.5)). Assume that an output unit connects only to itself, not to other output units (i.e., $w_{ii} \neq 0$, but $w_{ij} = 0$ if $i \neq j$). In such a case, typically $w_{ij} < 0$, and this parameter is interpreted as a leakage parameter: Activation gradually leaks from the unit.

4.2 Argue why any solution ($\gamma/C, \mathbf{w}C$) for any value of C, is just as good in equation (4.2).

4.3 Derive equation (4.6) from equation (4.5).

4.4* Derive the cross-entropy update rule if the potential hypotheses t_i are mutually exclusive (as indeed is plausible in our pet example; an animal is either a cat or dog, not both).

58   Chapter 4

The Geometry of the Delta Rule

Linear Independence

The MSE and cross-entropy learning rules are guaranteed to learn a task whenever
input vectors **x** are linearly independent. This is a much more relaxed situation than
the orthonormality assumption that we had to invoke in the Hebbian learning case in
chapter 3. To get an intuition about why this is the case, consider the following argument: Suppose that you have two empty vectors and you fill them with random numbers. The resulting vectors are almost surely linearly independent; but they are almost
surely not orthonormal.

However, linear independence is still quite limited in practice because if the number
of input vectors grows (beyond the tiny $n = 2$ example just given), linear independence
becomes more and more rare. For example, suppose that the vectors are two-dimensional
(2D), and consider the still relatively simple situation of $n = 3$ input vectors; here, one can
already be sure that the three vectors are not linearly independent. Thus, even with three
input vectors, it is already no longer guaranteed that one can learn the corresponding
input-output mapping. For this reason, linear independence is an extremely strict condition. I turn to a more feasible condition in the next subsection.

Linear Separability

If one has linearly dependent input vectors, a further condition must be checked to
determine if the input-output mapping can be learned at all. This condition is called
linear separability. Note that the issue is not whether an actual learning rule finds the
mapping. We are only considering whether a set of weights exists that implements the
input-output mapping. If the answer is negative, the learning rule will never be able to
find the weights, however sophisticated the learning rule is.

To explain linear separability, consider the input vector space, depicted in figure 4.3a.
In this space, each dimension corresponds to one dimension of the input vectors (hence,
in the example, the input vectors are 2D). Each input vector can be represented as one
point in this space. The dots and plus signs in figure 4.3a represent individual input
vectors (data points).

Let's assume that the activation function $f(.)$ is the threshold function (with a threshold of zero for simplicity); see figure 4.2b for an illustration. It is formally defined as

$$f(in) = \begin{cases} 1 \text{ if } in > 0 \\ 0 \text{ if } in \leq 0. \end{cases}$$

Thus, all points (x_1, x_2) that activate the output unit have $f(in) = 1$; and all the other
points have $f(in) = 0$. Because of the linearity principle, the points (x_1, x_2) that are
exactly on the boundary between the two categories satisfy the following relation:

$$in = x_1 w_1 + x_2 w_2 = 0. \tag{4.9}$$

Thus, those points (x_1, x_2) form a linear curve in the input space (see figure 4.3a).

The Delta Rule   59

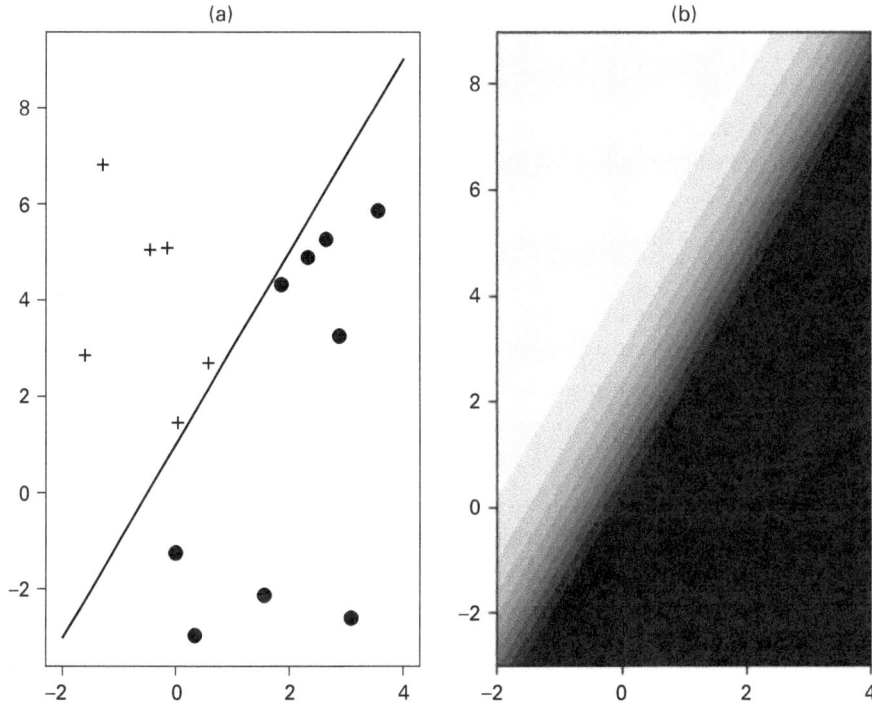

Figure 4.3
The geometry of the delta rule: (a) threshold activation function in 2D; (b) logistic activation function in 2D.

This provides an immediate intuition about which input-output mappings in a two-layer model are learnable and which are not: those mappings in which a line [through the origin $(x_1, x_2) = (0,0)$] exists, such that all category labels $(t = 0, 1)$ are on the same side of the line, are learnable by a two-layer network. For example, suppose that in figure 4.3a, all dots are from one category and all pluses from another category; in such a case, this mapping is learnable. Those mappings for which no such line exists are not learnable by such a network. We say that a mapping problem is linearly separable in the first case, and not linearly separable in the second case. Note that this makes it very convenient to check whether a problem is linearly separable, for the simple reason that our visual system is excellent in checking whether a line can be drawn between two 2D clouds of dots.

This visualization also provides an intuition for the earlier claim that mappings with linearly independent input vectors are always learnable by a two-layer network. In 2D space, two linearly independent vectors are two nonoverlapping points in space, and it's always possible to draw a line between them. But if a third vector is added, it is no longer guaranteed that the problem is learnable, and the linear separability test must be invoked to know whether a problem is solvable. This generalizes to more dimensions: In J-dimensional space, one can have at most J linearly independent input vectors;

typically if the number of input vectors is J or less, the vectors are indeed linearly independent and the mapping is learnable using the delta rule. But if the number of input vectors exceeds J, the linear separability test is required.

If a problem (such as the "cats versus dogs" classification) exhibits linear separability, the delta rule will typically find the optimum because the aim of the delta rule is to minimize error. Note also that linear independence and linear separability are properties of the *problem* [i.e., the combination of (input, output) vectors], not the algorithm that one uses to solve that problem.

Going from a requirement of orthonormality (Hebbian learning) to linear separability (delta rule) renders the delta learning rule much more powerful than the Hebbian learning rule. For example, in chapter 3 we saw that a Hebbian learning rule has difficulty storing even a few nonorthogonal digitized handwritten digits. Using the chapter 4 code on the GitHub repository (e.g., ch4_tf_logreg.py), the reader can check that learning this full MNIST data set is no problem for the delta rule. It can learn the patterns and generalize to novel patterns that were not presented during training.

To explain the MSE minimization rule, I treated the threshold as just another parameter in the model. However, in the geometric intuition that I am building here, it makes sense *not* to treat it as an arbitrary parameter, but instead to explicitly add it to the input. Thus, the line separating categories $y = 0$ and 1 becomes

$$x_1 w_1 + x_2 w_2 + \theta = 0. \tag{4.10}$$

This extra parameter allows the separating line to get out of the origin (0, 0). See figure 4.4 for an illustration of what the threshold θ represents in a one-dimensional

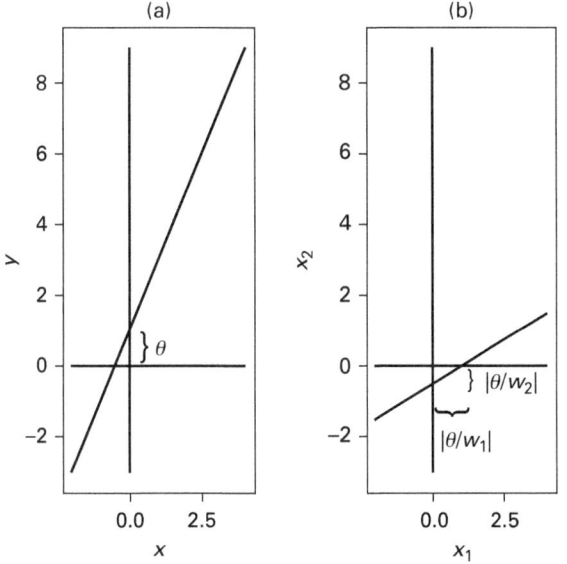

Figure 4.4
(a) Threshold illustration in a 1D function; (b) threshold illustration in a 2D function.

The Delta Rule   61

(1D; panel a) and in a 2D (panel b) function.[1] Keep in mind, however, that this geometric argument is just a way of shaping the problem to our visual and cognitive systems, and thus to make it easier to understand. Formally, parameter θ is still treated as a parameter by the (e.g., MSE minimization) algorithm, just like all the other parameters (the weights w_{ij}).

To make this geometrical argument, I had to invoke the assumption that $f(.)$ is a threshold function. Strictly speaking, the geometric argument applies only then. However, we can also make it work approximately to smooth logistic activation functions (as shown in figure 4.2c). Consider figure 4.3b. Here, we can again consider all points that conform to an equation like equation (4.10). In this case, these are all the points that satisfy $f(.) = 0.5$ (see the grayish zone in figure 4.3b). Points on one side of the line have $f(.) < .5$; points on the other side have $f(.) > .5$. Thus, this line again separates the two categories, but in a soft way. As one gradually moves away from the $f(.) = .5$ line (or the $f(.) = .5$ plane in three dimensions (3D); or the $f(.) = .5$ hyperplane in more than 3D), activation gradually decreases (or increases, depending on the side of the line; see figure 4.3b).

As noted before, the logistic function is the soft version of threshold function, and here one sees its multidimensional instantiation. One also can note that softness isn't always a disadvantage: it allows us to apply the gradient descent algorithm. Thus, the soft-threshold logistic function combines the best of two worlds: on the one hand, it allows the application of a powerful gradient descent algorithm; but on the other, it also allows a geometric intuition of what the algorithm is doing.

Exercises

4.5   In equation (4.9), which points (x_1, x_2) lead to $y = 0$, and which to $y = 1$?

4.6   With the logistic threshold function, why do points (x_1, x_2) with $f(in) = .5$ satisfy a linear relationship?

4.7* Demonstrate why linearly independent input vectors always allow you to reconstruct any desired target values t. (Hint: Consider it as a linear algebra problem $\mathbf{X}\mathbf{W}^\mathsf{T} = \mathbf{t}$.)

4.8   The Python machine learning library scikit-learn implements several algorithms for learning linear and nonlinear mappings (many more than could be discussed here). Use this library to fit a two-layer model with the OR, AND, and exclusive-OR (XOR) rules.

4.9   The Python machine learning library TensorFlow implements several algorithms for optimizing a goal function (many more than could be discussed here). Use this library to fit a two-layer model with the OR, AND, and XOR rules.

The Delta Rule in Cognitive Science

Prediction Error

In chapter 3, we mentioned that the Hebbian learning rule is a building block of other learning rules. We can see this for the MSE rule; it is basically the product of a term corresponding to the sending unit (x_j) and a term corresponding to the receiving cell $((t_i - y_i)\, y_i\, (1 - y_i))$. If we ignore the technical term $y_i(1 - y_i)$, the output activation term

t_i from the Hebbian rule has been replaced by the term $t_i - y_i$. So it is crucial in this rule that a difference is computed or a comparison is made. Actually, it will turn out that many learning rules have these two properties: (1) They consist of a factor from the sending unit and a factor of the receiving unit; and (2) they compute a difference between a prediction (y) from inside the network to a target (t). Learning consists of changing the weights in such a way that the prediction is as close as possible to the target.

The difference term is often called a *prediction error*; it is the difference between what is obtained (here, t) and what is predicted (here, y). In fact, this difference term is the reason why we call this the delta rule; as mentioned in chapter 1, *delta* is the Greek version of the letter d, the first letter of the word difference. One of the first prediction error–based learning rules was proposed by Frank Rosenblatt (1958) about half a century ago; this model had a threshold activation function (see figure 4.2b). A few years later, a delta rule with a linear activation function was proposed (Widrow & Hoff, 1960).

Another notable learning rule was developed a few years later by Rescorla and Wagner (1972), who constructed their famous model because they were interested in explaining blocking data from the animal learning literature (Kamin, 1969). In a blocking experiment, one initially combines stimulus A (say, a light) with an unconditional stimulus (say, a shock). Subsequently, a second stimulus called B (say, a tone) is added to A, again followed by the unconditional stimulus. Finally, stimulus B is presented on its own. In this case, it is observed that stimulus B does not predict the unconditional stimulus. This can be measured, for example, by testing whether the animal freezes in fear of expecting a shock. It is said that the stimulus A has "blocked" learning of the B-shock association. This phenomenon is accounted for naturally in a prediction error framework (can you see why?). More recently, neurobiological and causal evidence was found for the prediction error interpretation of blocking (Steinberg et al., 2013). Using rodent subjects, these authors stimulated the dopaminergic midbrain (which is thought to generate prediction errors in their specific experimental setup) during the compound (A and B together) presentation. Consistent with a prediction error account of blocking, they observed that this stimulation largely abolished blocking.

The Rescorla-Wagner model cannot account for several other findings in the conditioning literature (e.g., Kruschke & Blair, 2000). Moreover, the basic phenomenon of blocking is not always observed, especially in human subjects (Maes et al., 2016). Yet prediction errors as implemented in the delta rule play an important role in both in procedural (Sevenster et al., 2013) and declarative (De Loof et al., 2018; Greve et al., 2017; Pine et al., 2018) learning and memory. Studying the effect of prediction errors in learning, including in educational settings, is likely to be an important avenue for future research (Howard-Jones & Jay, 2016).

Because of its combined simplicity and power, the delta rule and its successors (see chapter 5) have a rich history in the last decades of cognitive neuroscience. In the remainder of this chapter, I discuss a few cognitive models that implement the delta rule. In each case, the delta rule yields interesting psychological properties.

Psycholinguistics

In psycholinguistics, models trained by the delta rule and its variants have been particularly influential. I briefly discuss two tasks that have been extensively modeled with this learning rule, namely past-tense formation and word naming.

One early model considered how children learn the past tense of English verbs (Rumelhart & McClelland, 1986). The authors started from a number of basic developmental facts about the past tense, most notably the three stages that children go through when learning this skill. In the first stage, children know only the past tense of a handful of high-frequency words (e.g., "go-went," "walk-walked"). In the second stage, children *regularize* irregular words; for example, they produce "goed" as the past tense of "go." In the same stage, when given a novel verb (e.g., "rick"), they would be able to provide a regular past tense form ("ricked"). Data like these have been interpreted as meaning that children must somehow represent the rule for generating the past tense. Finally, in the third stage, children know both the regular and irregular past tenses.

To understand this phenomenon, Rumelhart and McClelland devised a two-layer model using a variant of the delta rule to generate the correct past tense at output when provided a verb at input. After training, it was found that the model proceeds roughly through the same three stages as children do. In particular, in the first stage, the model knows the past tense of only a handful of high-frequency words. In the second stage, it regularizes irregular words (and would thus produce something like "goed" as output when "go" is provided as input). Finally, in the third stage, the model learned the past tense of both regular and irregular words.

An important point to note is that the model was able to respond in a rule-following way (following the rule "past tense = stem + ed"), even though the rule was nowhere represented in the model. All knowledge of the model was distributed across its connections (what we called w_{ij}) between input and output units. We can consider these weights to be microfeatures: no single connection represents the past tense rule for any verb or verb class. In fact, in most cases, it is typically very hard to understand (let alone verbalize) what the role of a specific weight is in the final performance of the model. Generating the past tense for each individual verb, in this theory, is a collective, emergent property of the whole collection of microfeatures that are encoded in the model's weights.

This modeling paper was followed by a long debate in psycholinguistics about whether the mechanism for generating the past tense, and producing language more generally, can be more accurately characterized as "connection-based" (in line with the models discussed in this chapter) or instead as "rule-based" (McClelland & Patterson, 2002; Pinker & Prince, 1988; Pinker & Ullman, 2002; Plunkett & Marchman, 1993). One of the criticizing papers (Pinker & Ullman, 2002) proposed that instead of a large set of connections between input and output features, humans have two stores available for generating the past tense. The first is a large collection of words; the second is a rule-based past-tense generation device. It is therefore called the *words and rules theory*.

Depending on the input word, either the words store or the rules store is addressed for generating the past tense of the word.

One important advantage of the connection-based account is that it can naturally account for *quasi-regularities* in language (McClelland & Patterson, 2002). Indeed, in the English language, past-tense exceptions are never completely irregular. For example, the verbs ending in *eep*) obtain the (quasi-regular) past tense *ept* (e.g., weep-wept, keep-kept)). The model presented here can account for this fact because similar words support each other during learning. For example, if weights are moved in the direction to produce the past tense of weep (wept), the weights will also be well positioned to learn the past tense of keep (kept), simply because the two past tenses share many weights. Such quasi-regularities appear throughout language, and are hard to understand from a purely rule-based account.

The dust has not settled on the past-tense debate, and a detailed examination is beyond our current scope. For our purposes, it is relevant to see that modeling played an important part in the theoretical debate because it allowed showing, via computer simulation, that a set of assumptions (children learn and store connections between input and output features) can generate nonobvious consequences for empirical data, such as the three stages of past tense formation. Theoretical progress will require model development and comparison of models to data on both sides of the debate.

A second psycholinguistic task is word naming (e.g., based on pictures); here, the effects of priming and interference have been key targets for modeling (Oppenheim et al., 2010). On the one hand, repeating a word facilitates its naming (repetition priming). On the other hand, naming a set of semantically related words (e.g., car, boat, plane) is more difficult than naming a set of unrelated words. In particular, naming response times (RTs) increase with every subsequent word from the same category, even if these are separated by several unrelated words.

Oppenheim and colleagues proposed a two-layer neural network model where semantic features are mapped onto response units called *lexical units* (see figure 4.5). The model was trained using the delta rule [equation (4.7)]. Due to learning, connections between semantic features (input units) and their lexical units (output units), such as from vehicular to car, increased. Therefore, naming became faster across repetitions (repetition priming). But, due to the delta rule, presentation of a car also decreased the connection between (input unit) vehicular and (output unit) boat, slightly impairing the performance on boat next time a boat is presented. Indeed, generating the correct word "boat" is supported by the connection from vehicular to boat, which was weakened due to the learning of car. Thus, the authors could explain both priming and interference, and their detailed properties, using a single model and learning principle. For completeness, I note that (like elsewhere in this book) discussion of this particular model does not imply that it provides the ultimate explanation for this body of findings. Indeed, word priming and its generative mechanism constitute a topic of debate in psycholinguistics (Howard et al., 2006).

The authors also relate their findings to an effect in the memory literature called *retrieval-induced forgetting* (Anderson et al., 1997). This is the phenomenon that remembering

The Delta Rule   65

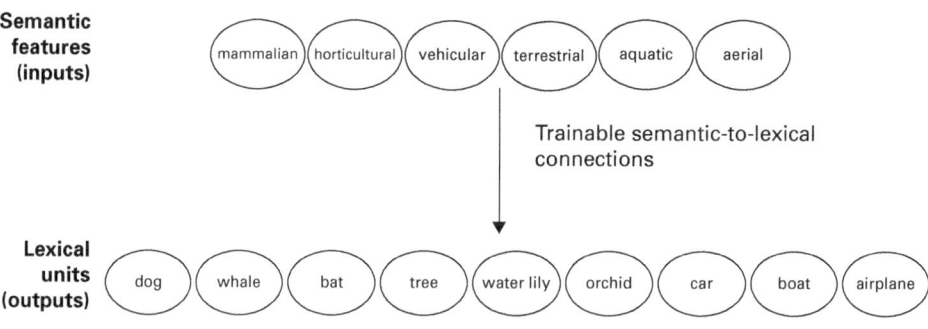

Figure 4.5
Word naming model. Semantic features map to lexical units, and the connections are trained with the delta rule. This figure was reproduced with permission from Oppenheim et al. (2010).

some information impairs later recall of semantically related information. Just as in the naming experiments from psycholinguistics, one may envision that a learning rule like the delta rule weakens connections between semantic information and memories as a result of recall of semantically related memories. It is interesting that the same principles can be applied across diverse fields, obviating the need for postulating different principles for each cognitive "function" (such as memory or language).

Numerical Cognition

Another example of the use of the delta rule in two-layer models comes from numerical cognition. Two celebrated effects in numerical cognition are the distance effect (already mentioned in chapter 2) and size effect. In the context of numerical cognition, it entails that symbolic numbers that are more distant, are compared more quickly and with greater accuracy (Moyer & Landauer, 1967). For example, it is easier to choose the larger of 2 and 9 than of 7 and 9. The size effect means that comparing larger single-digit numbers is harder; for example, it is easier to compare 2 and 3 than 8 and 9. The combination of these effects has typically been interpreted as originating from a "mental number line," a hypothesized left-to-right oriented continuum on which numbers are ordered (Dehaene, 1997; Izard & Dehaene, 2008). On this mental number line, larger numbers would be represented with less accuracy.

Some years ago, I trained a two-layer model using the delta rule on the number comparison task (and several other tasks; Verguts et al., 2005). The distance and size effects emerge naturally from delta-rule training if I presented numbers according to their naturally occurring frequency to the network (in particular, smaller numbers presented more often; Dehaene & Mehler, 1992). Postulating a compressed number line was not necessary to generate these effects. As often in computational modeling, a minimal set of assumptions accounted for more empirical data than would be considered possible without the actual computer implementation. This illustrates how modeling provides a conceptual tool to show which assumptions are needed to account for empirical data and which are not.

66   Chapter 4

The Rise, Fall, and Return of the Delta Rule

Despite its success in several domains, the delta rule also has some serious drawbacks. We saw this in the geometry of the delta rule (figures 4.3 and 4.4); a two-layer model can only construct linear boundaries. For some tasks, this is sufficient. Consider for example, the OR rule from logic (figure 4.6a). Here, an output is Yes (or 1) if either of the two input arguments is present, and No (or 0) otherwise. Consider the hypothetical rule "If a student is smart OR studies hard, he or she will pass the exam." If both conditions are fulfilled (i.e., a smart, hard-working student), he or she will pass the exam. Given that this problem is linearly separable (as can be seen in figure 4.6a), a delta rule can solve it. Similarly, consider an AND rule: "If a student is smart AND studies hard, he or she will pass the exam." Also in this case, the delta rule will be able to find weights that separate successful from unsuccessful students due to linear separability of this problem (see figure 4.6b).

For many tasks, however, this is not sufficient. This point was made forcefully in the 1960s in an influential book (Minsky & Papert, 1969). These authors demonstrated that a two-layer model (trained by the delta rule or similar algorithms) cannot solve (i.e., learn appropriate weights for) nonlinearly separable problems. They argued that nonlinearly separable problems are ubiquitous in daily life; an example is the XOR problem from logic (see figure 4.6c). For example, consider the rule "If you take the train XOR the car, you can go to the seaside." Obviously, and in contrast to the OR case, taking both the train and the car makes no sense (and will not get you to the seaside either). This is a task that humans easily figure out (you probably had no difficulty interpreting the XOR rule example), but a delta rule cannot learn it. The reason why is easy to see in figure 4.6c: there is no line separating the 0s from the 1s, and the XOR rule is therefore not linearly separable.

Models with more layers can solve the XOR problem, but no appropriate learning algorithm was known for such models at that time. Thus, Minsky and Papert (1969) argued, such models are not useful for understanding cognition. This led to what some

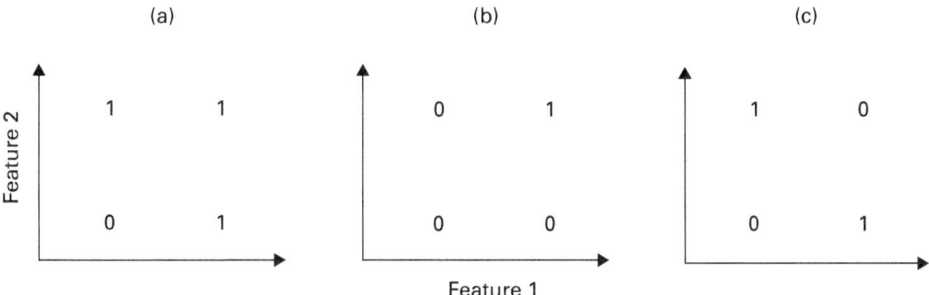

Figure 4.6
Geometric representation of three logical problems: (a) the linearly separable OR problem; (b) the linearly separable AND problem; (c) the linearly non-separable XOR problem.

The Delta Rule 67

have called "the Artificial Intelligence winter" in the 1970s (although some researchers did keep the fire burning during that era, such as Grossberg, 1980). As we will see in chapter 5 and later ones in this book, several subsequent computational developments, up to the recent "deep learning" models (Mnih et al., 2015), are a response to this challenge.

To sum up, two-layer networks are convenient tools for cognitive neuroscience: they have a sensible error function and can be used to solve any linearly separable task using a simple and intuitive learning rule. Several cognitive problems can be formulated as linearly separable tasks, and in such cases, a comparison of model and human performance yields insights into both. Unfortunately, though, not all tasks are linearly separable. Chapter 5 reports how we can deal with nonlinearly separable tasks.

Exercises

4.10 Consider a Stroop task model with two input modules (for colors and words) and one output module (for responses). Each input unit represents one color or one word; each output unit represents one word. Subjects must respond to color only. Is this task linearly separable?

4.11 Consider the Stroop task where a different dimension (color or word) is relevant for different trials. A third input module that contains two input units (one for each task) signals which dimension is currently relevant. The response layer is the same. Is this task linearly separable?

5 Multilayer Networks

A grown-up is a child with layers on.
—Woody Harrelson

Geometric Intuition of the Multilayer Model

In chapter 4, we considered the simple and elegant delta rule, whose dynamics could be relatively well understood by using geometric intuition. However, it also had a major drawback: it could only be used to solve linearly separable tasks. Although this is a huge advantage relative to the Hebbian learning rule, it remains too simple for modeling the complex cognitive tasks that humans (and other animals) are capable of. In the pet detector case, suppose that cats and dogs are distributed across two features, as in figure 5.1. With a two-layer (also called "shallow") model, we would not be able to tell the cats from the dogs. We need more sophisticated tools.

To address this, we extend our two-layer model by adding a hidden layer; see figure 5.2 for an illustration of this architecture. To make clear that each pair of layers can have its own activation function, I temporarily use the subscript and superscript indices x, y, and z for both activation functions (f) and connection weight matrices ($\mathbf{W}$). I again use the convention to index receiving units first. Activations then become

$$\mathbf{y} = f_{yx}(\mathbf{W}^{yx}\mathbf{x})$$

and

$$\mathbf{z} = f_{zy}(\mathbf{W}^{zy}\mathbf{y}) = f_{zy}(\mathbf{W}^{zy}f_{yx}(\mathbf{W}^{yx}\mathbf{x})).$$

The layer labeled y is the hidden layer (see figure 5.2). "Hidden" means that neither input nor supervisory feedback is injected into this layer. Conventionally, only the final (depicted as the rightmost) layer receives supervisory feedback (target variable t), and the hidden layers are placed between input (depicted as the leftmost) and output layers. To avoid clutter, I will usually remove the subscripts on the activation function f.

Figure 5.1
Geometric intuitions for multilayer models: cats and dogs in feature space.

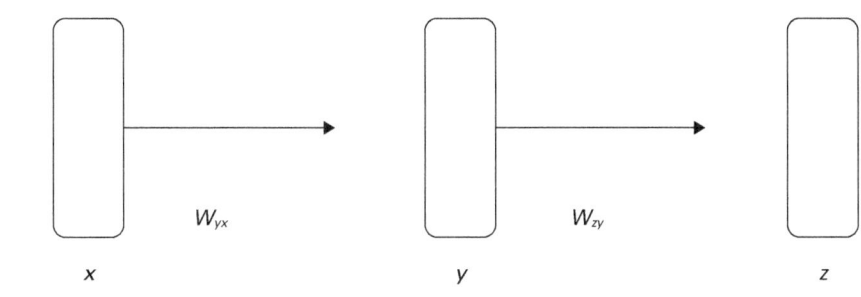

Figure 5.2
A generic three-layer network.

Models with at least one hidden layer of units are called multilayer models. When there are many hidden layers (with "many" not being precisely defined), they are called *deep models* (and training such models is called *deep learning*). Why are hidden units so powerful? To see this, we will use the same trick as in chapter 4: we first pump our intuition using the threshold activation function, and then "smooth out" the argument with the differentiable (and hence more powerful) logistic activation function.

Consider any hidden unit with a threshold activation function. Applying the knowledge from chapter 4, we can see that this hidden unit implements a linear separation of the input space. In particular, it separates the input space into two linear half-spaces, each of which is a convex set (see glossary). This holds for each of the hidden units. Now, at the output layer, several of these hidden units are linearly combined. As

Multilayer Networks 71

discussed in chapter 4, one such linear combination, when followed by a soft or hard threshold activation function, corresponds to taking a logical AND function across the hidden units. Taking an AND logical operation of half-spaces (each of which is represented by a hidden unit) results in another convex set. Thus, taking an AND function in the output layer across the units of the hidden layer will allow separating a convex set from its complement (see figure 5.3a).

This single hidden layer already increases the power of our model relative to the two-layer case. Indeed, we can now separate any convex set from its complement, whereas with a two-layer model, we could only separate a half-space from its complement. Now, suppose that we add yet another layer to the right in our model. Thus, we have four layers altogether, two of which are hidden. In the last (fourth) layer, we can combine the convex sets from the z-layer using an OR function. Perhaps surprisingly, this provides universal power (to any desired degree of approximation). Indeed, any set can be approximated by an OR-combination of convex sets (see figure 5.3b). For example, consider the cats and dogs from figure 5.1 again. The cats constitute two convex sets. The current argument shows that one can separate any of these two convex sets from its complement (using three layers); and that one can separate both convex sets (i.e., all the cats) from its complement (i.e., all the dogs) with one additional layer. Thus, with

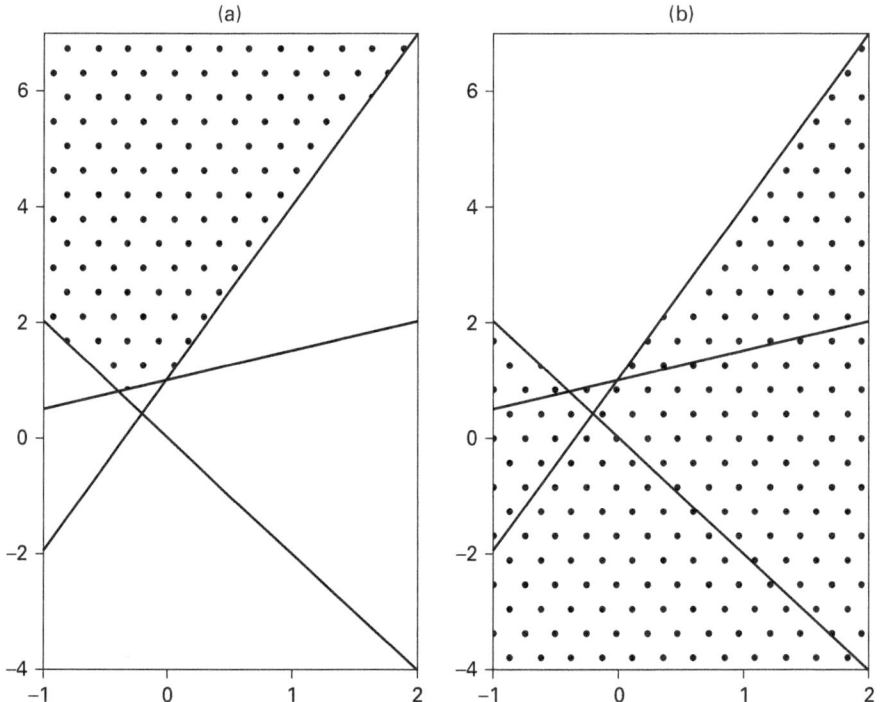

Figure 5.3
(a) An AND of linear functions (dots) is a convex set. Therefore, separating a convex set from its complement requires just three layers. (b) An OR of convex sets (dots). This construction allows for fitting very complex input-output mappings.

just two hidden layers (and four layers altogether), we can approximate the mapping in figure 5.1; and more generally, we can approximate any input-output mapping we desire.

If the activation function $f(.)$ is logistic, we can do even better: just one hidden layer suffices in this case to approximate any function. Demonstrating this fact is beyond the scope of this book; we refer the mathematically inclined reader to Hornik (1991) or to Bishop (1995). Recent deep neural networks typically have many more layers, sometimes even several hundreds of layers (LeCun et al., 2015; Silver et al., 2016), with all except the input and output layers being hidden (i.e., they do not receive supervisory target values). It is the case that with more layers, training a model is easier. That is, it is easier to find appropriate representations to solve a task. But strictly, all these many-layer networks could be replaced by networks with just a single hidden layer.

Exercises

5.1 Why doesn't it make sense to have a linear activation rule f_{yx} in the hidden layer?

5.2 Use the Python library scikit-learn to fit the XOR rule with a three-layer model. Compare your current result with that of exercise 4.8 in chapter 4.

5.3 Use the Python library TensorFlow to fit the XOR rule with a three-layer model. Compare your current result with that of exercise 4.9 in chapter 4.

Generalizing the Delta Rule: Backpropagation

As already noted by Minsky and Papert (1969) and discussed in chapter 4, it is not possible to apply the delta rule to a model with a hidden layer because there is no error term (delta) at the hidden layer. Again, by definition, a hidden layer is a layer that receives no outside information (either input or feedback). Fortunately, the gradient descent algorithm does not require this: application of the chain rule of calculus allows us to find the appropriate partial derivatives toward the hidden-unit weights.

Perhaps surprisingly, the derivatives toward the weights in the hidden layers (e.g., input to hidden weights) have an intuitive and relatively simple formulation in terms of the prediction errors of the weights in the subsequent layer (as explained in detail next). The realization that these resulting derivatives are so simple, and of similar form as the delta-rule derivatives, led to a widespread application of hidden-layer models in cognitive science, starting from the 1980s (Rumelhart et al., 1986b; Werbos, 1982) to its recent revival in deep learning in artificial intelligence (AI) (LeCun et al., 2015; Silver et al., 2016) and cognitive science (Annis et al., 2020; Kubilius et al., 2016; Shahnazian & Holroyd, 2017).

The learning algorithm works as follows. Consider again a mean square error (MSE) function that we aim to minimize for the generic model depicted in figure 5.2:

$$E = \frac{1}{I} \sum_{i=1}^{I} (t_i - z_i)^2. \tag{5.1}$$

Multilayer Networks 73

Note that E is now a function of weights contained in matrices $\mathbf{W}^{yx}$ and $\mathbf{W}^{zy}$. The back-propagation update rules for a three-layer network (figure 5.1b) are as follows for the hidden-to-output mapping:

$$\Delta w_{ij}^{zy} = \beta \delta_i y_j,$$

with

$$\delta_i = (t_i - z_i)\frac{d}{din_i}f(in_i).$$

This already looks pleasingly similar to the equation for the input-to-output layer in chapter 4. However, it becomes even nicer if we now look at the equation for the weights connecting the input layer to the hidden layer:

$$\Delta w_{ij}^{yx} = \beta \delta_i x_j,$$

with

$$\delta_i = \sum_k w_{ki}^{zy} \delta_k \frac{d}{din_i}f(in_i), \tag{5.2}$$

where k indexes across all the output units. Note that again, an update is of the same type as in the delta rule: there is a term relevant to the sending unit (x_j) and a prediction error term for the receiving unit (δ_i).

Let's take a closer look at the prediction error term for the input-to-hidden mapping in equation (5.2). The prediction error for hidden unit i in this case is a weighted combination (weighted by w_{ki}^{zy}) of all higher-level prediction errors that unit i projects to. This makes intuitive sense: if prediction error δ_k is large *and* unit i has strongly contributed to that prediction error (i.e., large w_{ki}^{zy}), then that unit i must be "informed" of this, and this will increase its own prediction error δ_i. However, if either the prediction error is small or unit i has not contributed to it (i.e., small w_{ki}^{zy}), then unit i should not increase its own prediction error. This is why we multiply w_{ki}^{zy} and δ_k. Finally, we add all the contributions across all output units.

This argument illustrates the broader issue of *credit assignment* in neural networks, which refers to how credit must be distributed across several units. Informally, the learning algorithm must determine which units performed correctly (or not) and which other units should receive the credit (or blame) for that. In the two-layer case, this credit assignment process was computationally relatively easy because each output unit receives its own tailored feedback. Because this credit assignment process is relatively easy in two-layer models, learning the correct weights is relatively fast in that case. But if units do not receive such tailored feedback, as in the hidden layers of a multilayer model, credit assignment is more difficult.

Metaphorically, the model cannot know which hidden unit should be credited for a good response or blamed for a bad one. Consequently, learning will be much slower (i.e., take more trials) in models that contain hidden layers. This is especially the case

74   Chapter 5

for deeper layers, whose units are far removed from feedback. Appropriately assigning
credit to different actors (units) is a recurring issue in neural network modeling, and the
previous discussion shows the backpropagation solution for it.

Regularization

Equation (5.1) constitutes the basic error function that backpropagation aims to minimize, but the function can be adapted according to the modeler's desires. For example,
a popular option is to add an extra term $\lambda \sum w_{ij}^2$ to the right side of equation (5.1).
Because the final function must be as small as possible, the model will have to compro-
mise between making the error small and keeping the weights small. Parameter λ is a
scaling parameter that determines the relative importance of the two factors: if $\lambda = 0$,
weight size does not matter; if λ goes to infinity, the model will only care about making
the weights small (in which case the best solution is to make all the weights zero). This
procedure will enforce a sparse (i.e., small weights) solution. This is often a good idea
because the optimal weight solution may not be unique (different weight configura-
tions lead to the same minimal error[1]), and adding an extra term makes sure that one
finds a solution that fulfills a useful property (here, weights as small as possible). This
has several advantages, including interpretability of the final weight configuration and
generalization toward novel stimuli (Kim et al., 2013). The addition of extra terms to an
error function (or goal function more generally), is called *regularization*; besides keeping
the weights small, several other goals can be enforced in this way (Kirkpatrick et al.,
2017; Quax & Van Gerven, 2018).

Some Drawbacks of Backpropagation

As noted, with a sufficient number of hidden units, any input-output mapping can be
implemented. But the backpropagation learning algorithm also has significant problems, some of which are described next.

Local Minima

If you have tried the exercises in chapter 4, you may have noticed that the delta rule
learning algorithm always ended up at the same solution (weight configuration). It
may not always actually solve the problem (e.g., if the problem was of the XOR variety
or another nonlinearly separable problem), but at least the algorithm always ended
up at the same solution. The reason for this is that the error function in a two-layer
network always has a nice shape; meaning that there is just a single minimum (see
figure 5.4a). Wherever you start the optimization process, gradient descent will always
bring you to that minimum. But with a hidden layer of units, this is no longer guaranteed. The error function may now exhibit local minima; that is, points where the gradient descent algorithm is attracted to, but that are not the actual minimum. Consider
figure 5.4b. This function has three local minima and one global minimum (i.e., four
minima altogether). If the optimization procedure starts close to a local minimum, the

Multilayer Networks   75

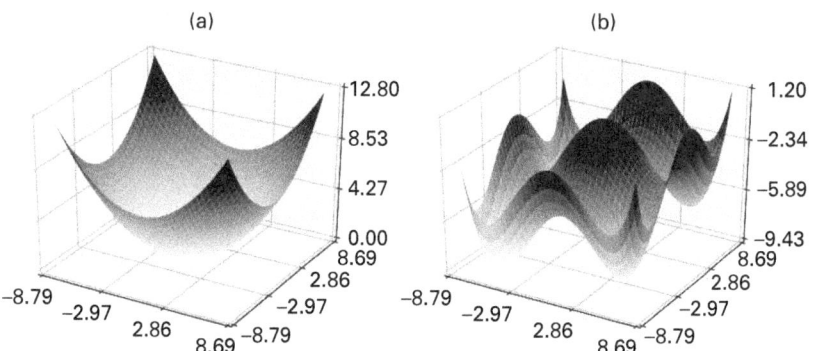

Figure 5.4
Global and local minima: (a) function with one (global) minimum; (b) function with several local minima.

gradient descent algorithm cannot help but slide down the hill toward the local minimum. The algorithm will converge, but not at the best possible solution. This is the problem of local minima (or local optima more generally).

Catastrophic Interference

Generally, weight changes in gradient descent learning are too global. In particular, when task B is learned after task A, all weights, including those that were nicely tailored for solving task A, tend to move in the direction that task B dictates. As a result, information from task A becomes unlearned. This problem is called *catastrophic interference* (McCloskey & Cohen, 1989), also known as the stability-plasticity dilemma. It is a major problem because it causes massive forgetting (of task A, in this example). Importantly for cognitive neuroscience, the problem also shows that standard backpropagation is not a viable model of how the human brain learns. Indeed, humans are not subject to catastrophic interference. For example, learning French first and then Russian may cause some mild impairment in the mastery of French; but learning Russian will not catastrophically wipe out all your knowledge of French. Catastrophic interference also occurs in the two-layer models discussed in chapter 4. However, there the problem is not as severe because the tasks that are given to two-layer models are typically quite easy (i.e., linearly separable tasks). As a result, the original, forgotten weights can quickly be learned again, and the problem of catastrophic interference is less severe. However, the more difficult tasks that most often confront multilayer models typically require many more learning trials, and unlearning of earlier information can lead to substantial inefficiency in this case. The biological brain fortunately does not suffer from catastrophic interference. There are two classes of theories on how natural evolution would have solved the problem of catastrophic interference. The first is replay: in this proposal, the information from task A would also be replayed when or after task B is being learned, during (awake and asleep) offline periods (McClelland et al., 1995; Mommenajad et al., 2018). The second is protection: Here, weights that

76   Chapter 5

are used in task A would be protected from overwriting while task B is being learned (Kirkpatrick et al., 2017; Verbeke & Verguts, 2019). Obviously, there is no reason why natural evolution would have had to make a choice, and both could well be true.

Vanishing and Exploding Prediction Errors

An error at layer n can always be backpropagated to the previous layer $n - 1$ according to a rule like equation (5.2); the process is illustrated in figure 5.5 in a four-layer model. However, prediction errors tend to either exponentially vanish (i.e., become extremely small) or explode (i.e., become extremely large) when they are backpropagated to deeper layers (Hochreiter & Schmidhuber, 1997). With vanishing prediction errors, the deeper layers do not receive the appropriate feedback; the final-layer feedback is only whispered to the initial layers (as it were). With exploding prediction errors, the solution becomes unstable; the final-layer feedback is shouted much too loudly to the initial layers (as it were). This is why training a model with many layers is computationally hard.

Varieties of Backpropagation

Several backpropagation algorithms have been developed over the last few decades. Some of these were built to address the computational problems mentioned in the previous section, or other such problems (Schmidhuber, 2014). Other models and algorithms were built because they addressed problems in cognitive neuroscience (O'Reilly, 2006). I cannot possibly give an overview of this huge literature; I suffice with noting a few developments that are relevant for cognitive neuroscience.

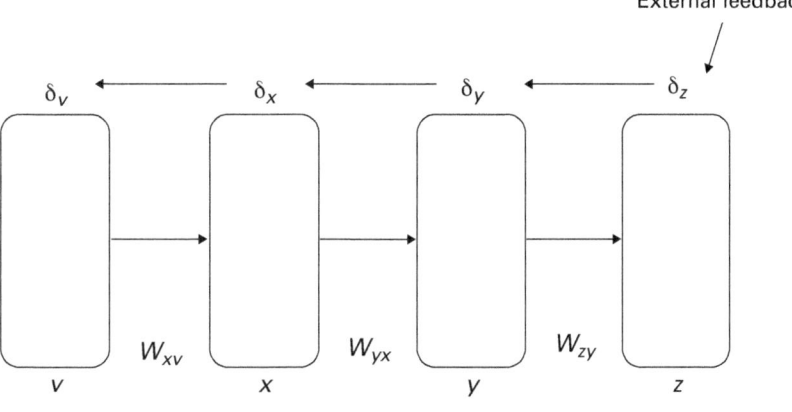

Figure 5.5
Backpropagating errors in a four-layer network model. External feedback is injected at the final layer z, from which the prediction error δ_z can be computed. Based on δ_z, the prediction error δ_y in layer y can be computed, and so on. Each transformation is roughly a multiplicative function, leading to either a vanishing gradient (if the multiplication constant is smaller than 1) or an exploding gradient (if the multiplication constant is larger than 1).

Multilayer Networks 77

Convolutional Networks

As mentioned, *deep learning* generally refers to learning with many hidden layers (and the more layers, the "deeper" the network is). One class of models that are typically "deep" are convolutional networks. A convolutional network is essentially a large (and deep) neural network model trained with backpropagation (LeCun et al., 1998, 2015). What makes the convolutional network special and powerful is that specific restrictions are placed on how the weights are allowed to change. In a huge network, there are typically very many (also unstable) local optima. By placing restrictions on the weights, the solutions become less frequent, more stable, and (sometimes) psychologically interpretable.

Consider the convolutional network picture shown in figure 5.6a. The input layer (the leftmost layer) provides the pixel input. The subsequent layer consists of a collection of maps (labeled C1: feature maps in figure 5.6a). Each unit in a feature map responds to a specific feature (e.g., a horizontal line). To construct a feature map, different units within the map should all respond to the same feature; the units in the map should differ only in their receptive field (i.e., what part of the visual field they respond to). For example, some units only respond to stimulation in the upper-left corner of the visual field, other units respond to stimulation in the upper-right corner, and so on, but all the units in the map would respond to a specific feature, such as a horizontal line. To accomplish this, the weight vectors projecting to each of the feature map units should all be the same. The fact that all the units in the map have the same incoming weights is known as *weight sharing*. This property is basically a strong restriction on which weights are allowed to change, in what direction, and at what time during learning. (It is possible to adapt the backpropagation algorithm to incorporate this weight-sharing restriction.) Because of this restriction, there are many fewer degrees of freedom for the gradient descent algorithm than in the unconstrained-weight case. As mentioned, such restrictions make the local optima problem much less severe.

Each map in this C1 layer responds to a particular feature. For example, one map could respond to a horizontal line, with each unit in the map having its own receptive field, another map to a vertical line (and again each unit with its own receptive field), yet another map to a specific color, and so on. Note that this approach is inspired by prominent attention theories from cognitive neuroscience (Treisman, 2006). Formally, each map can be considered as a linear filter for the property that it responds to. Such a function is formally equivalent to a mathematical convolution, hence the name *convolutional* layer (and by extension, the name of the network). As discussed earlier in this chapter (in exercise 5.1), the activation function $f(.)$ in each layer must be nonlinear and differentiable, such as the logistic one that we used in chapter 4. However, it is exactly this logistic function that causes the vanishing/exploding prediction error problem.

Other nonlinear and differentiable functions tend to have the same problem. Therefore, convolutional networks often use a rectified linear activation function (see glossary; such an activation function is also called a REctified Linear Unit, or RELU for short). In this activation function, the derivative is either 0 or 1; as a result, backpropagating prediction errors leads to backpropagated prediction errors multiplied

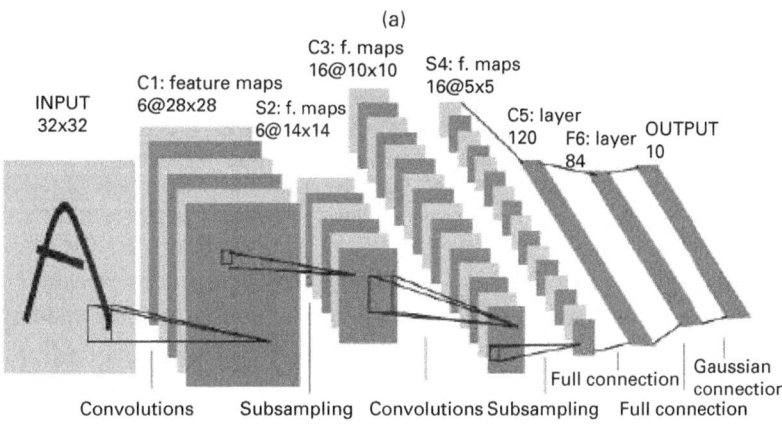

Figure 5.6
Convolutional neural networks: (a) model architecture. The leftmost layer provides the pixel input, the next layer contains feature maps, and the next layer implements subsampling. This two-step process (convolution–subsampling) is repeated, followed by all-to-all connectivity. (b) Classification accuracy on a picture by a convolutional network. Figure 5.6a was reproduced with permission from LeCun et al. (1998). Figure 5.6b was reproduced with permission from Ren et al. (2017).

Multilayer Networks

with either 0 or 1. Those prediction errors that are multiplied with 1 "survive" without vanishing or exploding. Strictly (and as noted in chapter 4), this approach is not mathematically correct because at point $x = 0$, the derivative of $f(x)$ does not exist. But in practical applications, this technical issue turns out to be outweighed by the practical advantage of not having vanishing/exploding prediction errors.

Each convolutional layer is typically followed by a *subsampling* layer (the first subsampling map is labeled S2 in figure 5.6a). In a subsampling layer, the average or maximum is taken from a number of adjacent units in the previous (i.e., convolutional) layer. Successive alternations of convolutional and subsampling layers then follow. Successive subsampling allows the network to gradually achieve translation invariance across successive layers. Stated otherwise, the units in C1 respond to only one specific location of the visual field; but due to the subsampling across several layers, the units in later layers respond to gradually larger parts of the visual field. The final few layers typically have all-to-all (full) connectivity (again, see figure 5.6a).

Convolutional networks are one of the reasons that neural networks have made a revival in the last decade, in both computer science and cognitive neuroscience. Early neural networks required isolated objects with handcrafted features as input for classification. Therefore, a human was needed to provide appropriate input to the network, rendering it not very useful for automatic computerized classification. However, convolutional networks are much more powerful and can classify images based on raw (pixel) input from realistic, cluttered scenes. An example is shown in figure 5.6b (Ren et al., 2016). This makes the convolutional network useful for automatized (e.g., online) classification in computer algorithms.

Recurrent Networks

Another interesting feature of multilayer networks is that the connectivity structure can contain loops. Thus, there may be a projection from layer A to layer B, as well as a projection back from layer B to layer A. A recurrent network can be trained with a variant of backpropagation called *backpropagation-through-time* (Rumelhart et al., 1986a).

Several recurrent networks have been proposed in cognitive neuroscience. One of the simplest such networks is the Simple Recurrent Network (SRN) (Elman, 1990). An SRN is a standard three-layer network (figure 5.2), but it has an extra context layer to which the hidden units project and which itself projects back to the hidden units. The context-to-hidden weights are also trained. Recurrence between the hidden and context layers allows the model to reprocess its older information, thus creating a working memory in the system, making them especially useful as models of cognition. The target for the network is the next input. For example, if one gives the network the sequence "A boy and a girl are walking" word by word, the target at "A" is "boy," at "boy" it is "and," and so on. Thus, the SRN learns to predict the next word in the sentence based on the previous input words.

Elman (1990) used the SRN to study language learning. He constructed a simple grammar and presented several sentences according to this grammar to the SRN. The author observed that the model could discover grammatical categories by merely seeing instances of grammatical sentences. For example, even though the network was never told explicitly, it discovered that there was a distinction between verbs (e.g., "see," "smell," "move," . . .) and nouns ("mouse," "cat," "dog," . . .). Within the category of nouns, it discovered that there was a distinction between animate versus inanimate nouns, and so on. This finding provides a proof of the principle that such concepts (like verbs and nouns) are not necessarily innate. A regularity extraction device (such as a neural network or, presumably, a human) can extract deep, statistical regularities during language processing, given a sufficiently rich data set. Of course, it also does not imply that these grammatical concepts are *not* innate, but it does show that the mere fact that humans can extract them without explicit instruction does not warrant a conclusion of innateness. Or, more specifically, it forces one to be specific about exactly what "innateness" means. For a more extensive discussion of this fact, as well as a discussion on some varieties of innateness, see Elman et al. (1996).

Another application of an SRN appears in the field of multistep decision making. Real-life tasks (unlike many laboratory tasks) typically consist of extended sequences of actions; they are multistep tasks intended to achieve a goal. An example of a multistep task would be making coffee, which consists of the steps "take coffeepot – put coffee in filter – add water—;" and so on. Suggestive of a hierarchical organization, each of these steps (or events) is again composed of different substeps (e.g., "put coffee in filter: open lid – take spoon – scoop ground"; and so on). An SRN was applied to multistep decision making by Botvinick and Plaut (2004). They trained an SRN on such multistep tasks and observed that its hidden units developed distributed representations of those task regularities.

Here, "distributed" is meant in the same way as discussed in the context of the past-tense network in chapter 4: single units do not necessarily contain the information for one specific piece of information, but jointly the units contain the information necessary to solve the task. This SRN does not contain any hierarchical representations, and goals do not explicitly drive behavior, although the model is of course goal-directed like every model described in this book—it aims to perform the task (obtain its goal) as well as possible. Despite this, the authors argued that the SRN could account for several empirical data on multistep decision making that did suggest the existence of explicit goals and hierarchy in human cognition, such as the fact that errors in multistep decisions tend to occur at event boundaries (e.g., between "take coffee pot" and "put coffee in filter," more than within events). The absence of explicit goals and hierarchy in this model was criticized by Cooper and Shallice (2006), arguing that both explicit goals and hierarchies are required to account for the full range of empirical data on multistep decision making. Following up on this, Cooper et al. (2014) built an adaptation of the SRN that contains explicit goals and hierarchical representations for multistep decision making.

Shahnazian and Holroyd (2017) also followed up on the Botvinick and Plaut (2004) model, and they proposed (for reasons beyond the current scope) that the anterior cingulate cortex (ACC) in the midfrontal cortex extracts the task regularities needed

Multilayer Networks   81

to perform such multistep tasks. A subsequent functional magnetic resonance imaging (fMRI) study with a multistep task paradigm suggests that ACC indeed contains distributed neural representations of the multistep task regularities, just as this SRN does (Holroyd et al., 2018).

Radial Basis Function Networks

One final type of network that I mention here is the radial basis function (RBF) network. Typically, an RBF network consists of three layers. The activation function at the hidden layer is not logistic, but rather it takes a radial form for each unit i:

$$f(\mathbf{x}) = \exp(-\|\mathbf{x} - \mathbf{p}_i\|), \tag{5.3}$$

where $\mathbf{p}_i$ is the "codebook vector" for hidden unit i; that is, the vector that the unit responds most vigorously to. One can consider the vector $\mathbf{p}_i$ as the vector of axons feeding into unit i, just as we did for vector $\mathbf{w}_i$. A hidden unit with a logistic activation function typically responds to half of the input space (see figure 4.3b in chapter 4). In contrast, a hidden unit with a radial activation function as in equation (5.3) responds to only a limited part of the input space; see figure 5.7 for an illustration, where $\mathbf{p} = (1, 0)$. Figure 5.7 illustrates that this unit responds vigorously to input pattern (1, 0); but its response gradually becomes weaker as one moves away from this input pattern. Activation decays in a circular (or radial) shape—hence the name of this network. For more information on the formal properties of RBF networks, the reader can consult Bishop (1995).

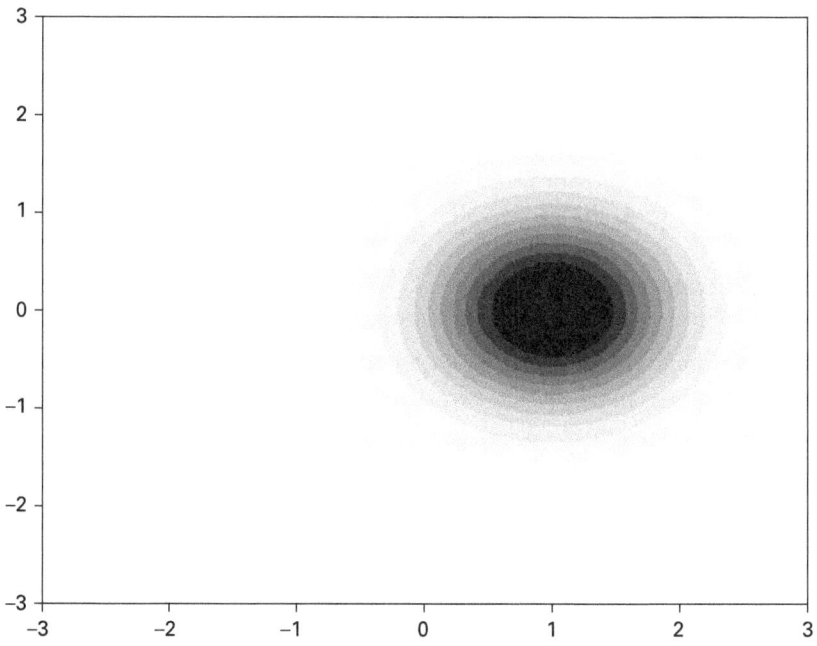

Figure 5.7
An RBF responds to just a limited part of the input space [here, centered on point (1,0)].

RBF networks have been used extensively in the study of categorization (Kruschke, 1992). One question that generated extensive debate is what exact information is stored for a specific category. For example, there may be just one prototype codebook $\mathbf{p}_{dog}$ to represent all dogs. Its corresponding unit in the hidden layer would then respond most strongly to the most prototypical dog. Another prototype codebook $\mathbf{p}_{cat}$ would then represent all cats and respond most strongly to the most prototypical cat. Alternatively, at the other end of the theoretical spectrum, each and every exemplar (i.e., each individual cat and dog) may be represented by its own codebook. For example, there could be a unit with codebook vector $\mathbf{p}_{Quirine}$; this node would then respond only whenever a specific dog (named Quirine) is presented. Pursuing this line of thought, there could even be a unit with codebook $\mathbf{p}_{Quirine, kitchen}$ that responds to Quirine only when she happens to be in the kitchen. In principle, humans could store the very general prototypes ($\mathbf{p}_{cat}$), the very specific individual exemplars ($\mathbf{p}_{Quirine, kitchen}$), or anything in between. A large body of categorization literature has investigated the extent to which humans store individual exemplars or instead more abstracted prototypes (Minda & Smith, 2011; Nosofsky et al., 2018).

Another domain where RBF networks have been extensively used is coordinate transformation in motor neuroscience. Suppose that a person sees a cup of coffee that he would like to grasp. For this to work smoothly, the grasping arm and the object must be placed in the same spatial coordinate frame. However, the cup is originally projected on the retina and thus in eye-centered coordinates, whereas the movement must be specified in joint-centered coordinates. Moreover, the eye itself can be looking left or right, and is thus specified in head-centered coordinates, Hence, several transformations must be carried out to go from information on the retina to an appropriate grasp, and RBF networks have proven very useful to carry out such transformations (Deneve & Pouget, 2003). Consequently, it has been hypothesized that the primate brain possesses RBF networks for coordinate transformation; there is indeed evidence for this conjecture (Avillac et al., 2005).

Networks and Statistical Models

The reader may have noted that there is a relationship between the networks discussed in chapter 4 and this chapter on the one hand, and traditional linear statistical models on the other. This relationship is indeed very close.

Consider, for example, a situation where we want to perform linear regression with independent variables X_1 and X_2, to predict the dependent variable Y. Typically, we put the full data matrix in a statistical package, which subsequently churns out the appropriate regression coefficients for X_1, X_2, and the intercept. Alternatively, we could present the same data to a two-layer network model. Rather than presenting all the data in a batch, we would now present to the network each data point separately. Each trial in a neural network context indeed corresponds to a data point in a standard regression context. Each input unit corresponds to one of the independent variables; the bias unit corresponds to the intercept; and the output unit corresponds to the dependent variable. The regression coefficients that come out of the statistical analysis turn out to be exactly the same (up to the measurement error, of course) as the weights that one would obtain from a two-layer network after training. If the activation function is linear, the network performs linear regression; if the activation function is logistic, as in equation (4.3), it performs logistic regression (Bishop, 1995).

Because the trials (data points) are presented one by one to a network rather than in a single batch as one does in a statistical model, the two-layer model is said to perform online linear (or logistic) regression. For the same reason, the delta rule performs stochastic gradient descent; stochastic because a random (stochastic) process samples a trial, and estimates the gradient of the MSE error function on that trial.

Just like a two-layer model implements online linear (or logistic) regression, models with hidden layers implement online nonlinear regression. Indeed, if the activation function in the hidden layer is nonlinear, then the hidden units construct a nonlinear combination of the original input variables (e.g., X_1, X_2). This nonlinear combination then constitutes a new variable that maps to the dependent variable (Y).

Despite their formal similarity, it is important to note that the concept of goal optimization fulfills a very different purpose in the two contexts. In the modeling context of most of this book (except in chapters 6 and 7), the goal is considered to be of the organism. The multiple trials mimic the organism's learning trajectory to reach that goal. We may inspect this trajectory and evaluate whether it corresponds in important respects to empirical data (e.g., Mayor & Plunkett, 2010). If it does, we consider this another argument supporting the idea that the model is cognitively plausible. In contrast, there is no reason whatsoever to present the data across multiple trials in the statistical linear regression case. In fact, that would definitely be suboptimal: it introduces an extra source of noise, whereas the goal can be optimized in a single sweep through the data. Stated otherwise, in the statistical context, the goal of optimization is not a goal of the model itself; rather, it resides in the mind of the scientist, who wants to obtain the best regression coefficient estimates. How we get to these best regression coefficients is irrelevant.

Multilayer Networks in Cognitive Science: The Case of Semantic Cognition

Because of their representational power, backpropagation models have been very popular in cognitive science. As is often the case, one important use is demonstrating that two apparently very different theoretical positions can be reconciled. I will shortly describe an example of this approach in a field that is closely related to the human categorization literature mentioned earlier in this chapter (namely, semantic cognition).

The term "semantics" refers to meaning, and the field of semantic cognition studies how humans learn, store, and process such meanings. Here, two broad theoretical views have been proposed. The first considers semantic cognition from a similarity-based perspective. Similarity theorists argue that objects are semantically categorized depending on their feature-based similarity to each other (Nosofsky et al., 2018). Despite their mutual disputes, the prototype and exemplar theories of categorization discussed earlier in this chapter and the feature-based models discussed in chapters 2 and 3 are all similarity-based theories. In this approach, any object consists of a feature bundle, and the more similar the two bundles, the higher is the probability that the two objects are categorized as belonging to the same semantic category. For example, the feature bundles of the sparrow and the robin are likely very similar (at least in a nonexpert mind), and therefore these two animals are considered similar.

A very different approach toward semantic cognition is the theory-theory approach, which argues that similarity does not determine category membership. Theory-theorists would refer, for example, to a category of birds, consisting of such physically dissimilar entities as eagles, robins, penguins, and chickens. In the nonnatural world, differences between entities of the same category may be even larger. Consider, for example, the diversity of things that can function as a stop sign on the road (a policewoman, a traffic light, a traffic sign, . . .). It is hard to see which feature bundle holds this category together. Instead, so theory-theorists would argue, categories are bound together by a theory that the subject holds (hence its name) about the category (Murphy & Medin, 1985). For example, a theory about birds would hold that they fly, have light bones, have feathers, and so on. In this view, this theory is what keeps the category together, not feature-based similarity between individual birds (or stop signs). When queried, a person who understands the concept of "bird" would likely confirm that a bird does not speak, has no wheels, does not attend opera, and so on. This is not because these properties belong to the bird feature bundle; it is because the concept of bird is held together by a theory based on which the person can reason about these animals.

Rogers and McClelland (2004) have argued that these two views can be reconciled. A key part of their argument was the fact that features in the real world show a pattern of *coherent coactivation*. In birds, for example, the features "fly," "light bones," "feathers," and others typically cooccur; and the same features rarely occur in combination with "having wheels" or "attending operas." The authors argued that a multilayer network, which naturally computes similarities between its inputs, would pick up on this coherent covariation, and thus show the properties that the theory-theory would predict. In this way, a similarity-computing device such as a neural network would also exhibit theory-theory properties.

To demonstrate this, they trained a multilayer model on a number of facts about living things (e.g., a salmon can swim, a robin is a bird, and so on). After training, they observed that the living things naturally clustered in high-dimensional, hidden-unit spaces (see figure 5.8). In line with theory-theory, the features that are most category-relevant (e.g., can fly, has feathers) drive this clustering of animals in the model. The clustering is not just based on overall feature-based similarity. For example, the feature "is yellow" may be shared between a lemon and a canary; but it is a superficial feature, not supported by any theory about birds, and so it does not contribute to the very low similarity of lemons and canaries in the model. However, in line with the similarity-based approach being a neural network model, similarity is still feature-based in the model.

Multilayer Networks 85

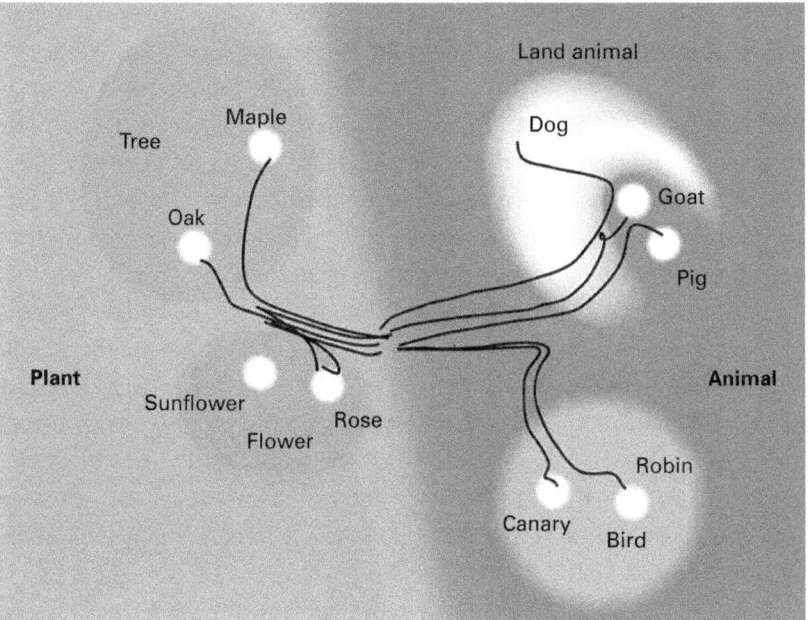

Figure 5.8
Internal (hidden unit) representations after training the semantic cognition model. This figure is reproduced with permission from McClelland and Rogers (2003).

In addition to reconciling the two broad views on semantic cognition, the Rogers and McClelland (2004) model accounts for a host of empirical data. For example, children tend to first learn broad categories (e.g., animals versus plants) and only later fine-tune the more specific categories (e.g., canary versus robin). This happens in the model too. The trajectory data (black lines) in figure 5.8 demonstrate that initially, only broad categories can be distinguished. With repeated exposure, there is a gradual differentiation of the categories.

Also, in Alzheimer's dementia, categories are again unlearned in the reverse (fine-to-broad) direction (initially, canaries are confused with robins; and only later are animals confused with plants or inanimate objects). This opposite trend is also observed in the model. One can intuitively see this by imagining that some noise is injected into the semantic space of figure 5.8. A little bit of noise may lead to confusion between "canary" and "robin," but not between "canary" and "oak." Only when a lot of noise is injected (presumably corresponding to severe brain damage) would broad categories be confused.

Criticisms of Neural Networks

In the last several decades, neural network models like those discussed in chapter 4 and this chapter have been instrumental in shaping how scientists think about cognition. From the start, however, they have also met with several criticisms. One criticism of deep neural networks is that little is known about what such networks do and how they do it (McCloskey, 1991). However, recent research is actively discovering the mathematical principles of deep (and not so deep) networks. Concerning the semantic cognition studies, Saxe et al. (2019) trained a three-layer model (but only with linear activation functions) on this task. Their formal analysis demonstrated exactly why the model learns in a hierarchical fashion, in the sense that it first extracts the overall mean, then the animal versus plants distinction, then the bird versus fish distinction, and so on. However (as the astute reader will note), didn't exercise 5.1 demonstrate that it makes no sense to have a linear-only deep learning model? Actually, what exercise 5.1 demonstrates is that for the final weight solution, it doesn't matter whether one has two or more layers. Saxe et al. (2019) demonstrated that for the *trajectory* toward that solution, adding deep layers can be of value nevertheless. In fact, they also demonstrated that this hierarchical development does not occur in two-layer networks. A related criticism is that neural networks are just a kind of (non)linear regression. As we have seen in this chapter, this is true. But in this line of reasoning, one might also argue that a computer just implements arithmetic. What matters is what you do with these simple building blocks.

The previous criticism is that neural networks are in some sense too strong because they can, like a black box, fit any data. Another criticism is that they are in another sense too weak because they often have trouble generalizing to novel (nontrained) data. Humans can generalize very quickly from earlier experience, but neural networks typically do much worse than humans in this respect (Lake et al., 2017; Marcus, 2006; a recent nontechnical exposition of this argument appears in Marcus & Davis, 2019). The reader can easily check this using the TensorFlow Python files in the chapter 5 directory (e.g., the file ch5_tf_imag_classif_3layer.py). Here, a distinction is made between error (after training) on the trained data set versus error (after training) on a data set that was held out during training (test data). Whereas it is typically easy to obtain very high accuracy and very low error on the training data, this is definitely not the case for test data, especially for more difficult tasks. Performance on test data is, however, critical to know whether any (natural or artificial) agent can perform a task. Suppose an agent can correctly say "cat" or "dog" to the trained cat and dog pictures on which he was trained. This is clearly not enough to know whether the agent has really extracted the knowledge of what cats and dogs are. For the latter, we require that the agent can also label novel (i.e., test) cats and dogs correctly. Hence, making neural networks generalize to novel data is an important challenge for them.

One important reason for this problem is that standard neural networks do not sufficiently exploit *compositionality*, which means that information is stored in a modular (i.e., decomposed) fashion (Fodor & Pylyshyn, 1988). For example, suppose that one learns a family tree containing pieces of information like "Bob is the father of Mary." If the information about this specific family and the general concept of a family tree are smeared out (i.e., distributed) across the many units of the network (as a standard neural network would do), then it is hard to generalize the learned information to a completely different family (even if it has the same family structure). However, if a model would learn about family trees in general, and separately learn about any specific family, then the former type of information about family trees in general might be more easily applied and generalized to a novel family. Compositionality is an active area of research in cognitive modeling (Lake et al., 2017; Schulz et al., 2017), including in neural networks (Kim et al., 2013; Whittington et al., 2020). One important instance of compositional representations is comprised of hierarchical representations; network models with hierarchical structure are also actively studied (Frank & Badre, 2012). Finally, if representations are compositional, they must also be bound together at some point; this is the binding problem in cognitive neuroscience (Treisman, 1996). Although a detailed explanation is beyond the current scope, I note that the binding problem is also an active area of research in neural networks. There is a number of theories on how the brain can solve this, including the use of neural oscillations for phase-locking distinct representations, thus allowing the representations to exchange information (Hummel & Holyoak, 2003; Verbeke & Verguts, 2019; von der Malsburg, 1995).

Another criticism is that it is hard to treat neural networks from a standard statistical perspective. In a standard statistical framework, a model is fitted and quantitative predictions can be derived from the resulting model (Ratcliff, 1978). Indeed, due to their complexity, fitting neural network models to empirical data is not trivial, and for this reason, neural network modelers often restrict themselves to demonstrating qualitative rather than quantitative fits to empirical data. This qualitative approach has advantages. Most important, if one is not forced to fit a specific data set, one often can account for a broader scope of data than is possible in a model fitting framework (e.g., Wang et al., 2018). Recall from chapter 1 that placing a data set or empirical finding in a common framework with other (seemingly unrelated) data is indeed an important aim of cognitive modeling. Also, the qualitative approach allows one to construct and investigate more complex models than would be possible in a statistical framework. Nevertheless, quantitative model fitting and comparison have major advantages too. One such advantage is that they allow for constructing and empirically comparing subtle variants of the same model (e.g., Daw et al., 2011). I consider statistical estimation and statistical model testing and comparison in chapters 6 and 7.

Finally, another common criticism is that the backpropagation learning rule is not biologically plausible. As discussed in chapter 1, whether this is a problem or not depends on the level of modeling that one is interested in. If we are interested in modeling semantic cognition, we do not want to include detailed cell membrane properties in our model units. However, the specific problematic issue for backpropagation that is often raised is that backpropagation assumes that feedforward and feed-backward weights are symmetric (i.e., $w_{ij} = w_{ji}$). But didn't we see that backpropagation is usually applied in feedforward models, where activation flows from left to right? Well, yes, but when we propagate error backward through the network [right to left, via the learning rule in equation (5.2)], we actually use the backward weights, and we indeed assume that $w_{ij} = w_{ji}$. Given that connections in biological neural networks between two neurons are typically not symmetric, and given that it's unclear how the network should manage the requirement that $w_{ij} = w_{ji}$, it is a valid point that the assumption of symmetry can be difficult to defend for biological networks.

Another formulation of this criticism is that the backpropagation learning rule from equation (5.2) is not local. Recall that when we discussed the general format of a learning rule in the context of equation (3.4), we said that a learning rule is often of the form "information about unit i multiplied with information about unit j." If the two pieces of information are not locally available to the respective unit, however, we say that the learning rule uses nonlocal information. Because a neuron can only use information in its local vicinity, a nonlocal learning rule is considered to be biologically implausible. Hence, if the weights w_{ij} are not equal to w_{ji}, then the learning rule uses nonlocally available information to compute its updates.

Recent work has also addressed this issue. It has been shown that just random backward weights can also support efficient learning—that is, simply replace the weights w_{ij} in equation (5.2) by random numbers (Lillicrap et al., 2016), in an algorithm called *feedback alignment*. This may be very surprising: how could random weights support an appropriate distribution of error across several layers of units? The important point here is that the feedback weights are random, but also fixed. The network will thus attempt to make its feedforward matrix (which is free to change) approximate the feedback matrix (which is why this algorithm is called *feedback alignment*), while at the same time implementing the required input-output mapping. More generally, algorithms that approximate the computational efficiency of backpropagation, but with higher biological plausibility, constitute an active area of research in cognitive computational neuroscience (Lillicrap et al., 2020; Roelfsema & van Ooyen, 2005).[2]

To sum up, backpropagation is a flexible algorithm for training multilayer neural networks. Its goal consists of minimizing an error function at the output layer. Backpropagation is used in cognitive neuroscience (as a model of the brain) and in AI (for solving various tasks) and forms an active area of research in both fields.

6   Estimating Parameters in Computational Models

The purpose of models is not to fit the data, but to sharpen the questions.
—Samuel Karlin

In the previous chapters of this book, I discussed some approaches to model building. However, just as for standard statistical models (e.g., the linear model), statistical methods can be used to estimate the parameters of and empirically compare such methods, called *statistical model analysis*. This forms the topic of this chapter and chapter 7. I refer the reader who wants to know more details about statistical model analysis to Farrell and Lewandowsky (2018).

I have discussed how activation rules determine changes in unit activations (x_i) and how learning rules determine changes in network weights (parameters w_{ij}). Activation rules and learning rules were set up in such a way that activations and weights were changed in order to optimize a goal function. Other parameters remained fixed. For example, in all learning models, there was a learning rate β, which was considered to be fixed. But the network architecture itself (e.g., the number of layers or number of units within a layer) yields several discrete parameters that must be fixed. Slope γ was another fixed parameter; and in the diffusion model, the drift rate and boundary parameters also remained fixed throughout an experiment.

How do we know what values these parameters must take? There are two basic approaches to this issue. The first is to explore at least part of the parameter space and evaluate how the model behaves in various parts of the parameter space. This approach is briefly discussed in the next section. The bulk of this chapter, however, is about the second approach (namely, to estimate the parameter values based on empirical data, called *model estimation*).

Parameter Space Exploration

One can fix unknown parameters to some plausible value. A learning rate, for example, must be larger than 0 and smaller than 1 (or more precisely, smaller than a value above which the optimization process would diverge; see chapter 1, exercise 1.7). Within this range, the modeler can pick a parameter value semirandomly according to plausibility, aesthetics, or historical reasons. If you think that this seems a bit arbitrary, you are right. However, the basic idea behind this approach is that the exact value that such a parameter takes does not (or at least should not) really matter. To demonstrate that the parameter does not really matter, the model should yield qualitatively similar results, whatever its precise value. Thus, researchers can try to convince their readers by searching the parameter space and demonstrate that a large swath of the parameter space provides qualitatively similar results (Calderon et al., 2018; Rombouts et al., 2015). An example is shown in figure 6.1a from Rombouts et al. (2015). Here, the authors demonstrated that whatever their network size (on the x-axis), the model nearly always converged (i.e., learned the task). Hence, the exact parameter values (network sizes) that were chosen (at least within the studied range, and for the chosen, fixed, learning rate) did not really matter.

Another exploration approach is parameter space partitioning; here, one acknowledges that different regions of the parameter space may lead to qualitatively different results, but one catalogs exhaustively how various parts of the parameter space behave (Pitt et al., 2006; Steegen et al., 2017). See figure 6.1b for an example of different regions of the parameter space predicting different relative difficulties of stimuli, labeled 1-5. For example, in the largest part of the parameter space, the model predicts that stimulus 5 is the easiest, then 1, then 2, and so on.

Exploring the parameter space quickly becomes infeasible in more dimensions, however. This problem is called the *curse of dimensionality*. Specifically, the size of the search

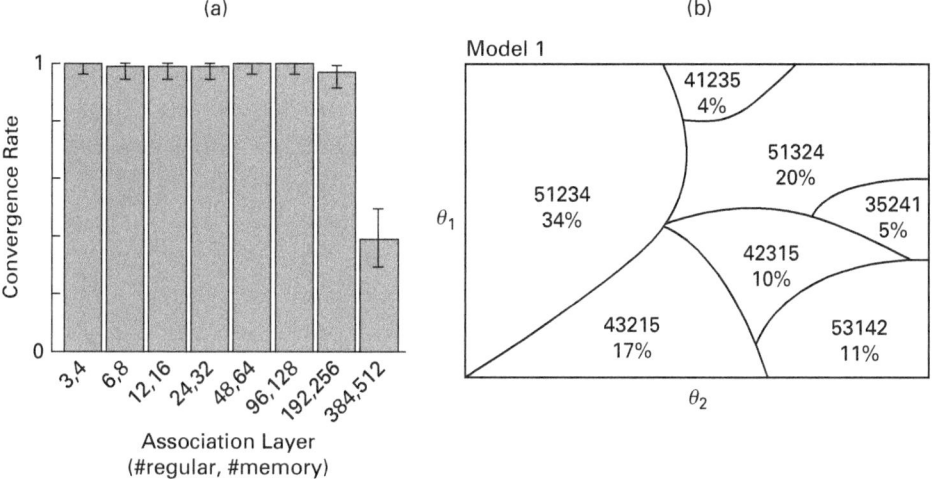

Figure 6.1
Exploring the parameter space: (a) the neural network converges for virtually every network size in the tested range; (b) different parts of the parameter space yield qualitatively different properties. Figure 6.1a was reproduced with permission from Rombouts et al. (2015). Figure 6.1b was reproduced with permission from Steegen et al. (2017).

Estimating Parameters in Computational Models 91

space grows exponentially with the number of parameters. For example, suppose that we have G parameter values in each dimension; for example, we may discretize the learning rate in the continuous space 0 to 1 into $G = 11$ points $(0, 0.1, \ldots 1)$; and there are D parameters. In this case, the total number of parameter values that needs to be searched is G^D. Clearly, this quickly becomes a very large search space. Perhaps for this reason, parameter space exploration is relatively rarely applied.

Parameter Estimation by Error Minimization

The second way to treat fixed parameters is to estimate them based on empirical data, just as we would, for example, estimate the regression coefficient parameters in a linear regression model. In effect, the cognitive model is treated as a statistical model whose parameters can be estimated.

To explain this approach, let's start with the familiar linear model from statistics:

$$Y(n) = \sum_{j=1}^{J} w_j X_j(n),$$

where $Y(n)$ could be the data of subject n on a specific (e.g., intelligence) test, which is explained by factors X_j (e.g., social status, education, nutritional variables, . . .), each with their own regression weight w_j.

The test scores $Y(n)$ of each subject n are known, as well as the factors X_j; they are the data. To find the optimal regression coefficients w_j, one can consider the following error function:

$$E = \sum_{n=1}^{N} \left(Y(n) - \sum_{j=1}^{J} w_j X_j(n) \right)^2, \tag{6.1}$$

and this quantity is minimized as a function of the parameters w_j. This approach of constructing an error function that needs to be minimized is formally similar to the approach from the previous chapters. However, it is important to see that the use of the error function in equation (6.1) is conceptually very different. When adapting activation (x) and weights (w) in the networks discussed in previous chapters, the goal function and its resulting algorithm were thought to correspond somehow to what a subject does. In parameter estimation, on the other hand, the goal function and its optimization are merely statistical tools to find the parameters that provide the best match to the empirical data. Stated differently, optimization can be considered from the modeler's perspective (this chapter) or from the model's perspective (e.g., chapters 2–5).

In the case of the linear model, the error function E is so simple that one can find the optimal parameters w_j analytically. In matrix notation, the parameters that minimize E in (6.1) are

$$\mathbf{w} = \left(\mathbf{X}^T \mathbf{X} \right)^{-1} \mathbf{X}^T \mathbf{Y}. \tag{6.2}$$

This is the result that you will get if you apply standard linear regression to the data in your statistical package of choice with the ordinary least squares method.

In ordinary least squares, simplicity comes at the price of making rather strong assumptions about the data. For example, this approach implicitly assumes that noise at each two different measurement units is uncorrelated. If we want to loosen this assumption, and thus also estimate the covariance structure of the noise, the analytical approach just discussed will no longer work. In such a case, we would have to apply a gradient descent for the linear model too (or another algorithm that starts at a random value and iteratively finds the optimal solution). Such iterative algorithms for complex linear models are standardly applied in, for example, the statistical analysis of functional magnetic resonance imaging (fMRI) data (Friston et al., 2002).

Note that regression coefficients are usually denoted by β_k in the (applied) statistics literature. I don't use this notation in order to avoid interference with the learning rate β from earlier chapters. I instead use w_k, to allow the reader to generalize easily from the weights notation in earlier chapters.

As intuitive as minimizing errors is, the approach does have some drawbacks. The exact choice of what the error function looks like is partially arbitrary. Why, for example, were the errors squared in equation (6.1), rather than raised to the fourth power? This question is not easy to answer. The most important problem with error minimization, however, is that the statistical properties of the estimates that come out of the analysis are not always well understood. To confront these issues, one typically starts the optimization from a slightly different standpoint than the one of minimizing error, discussed next.

Exercises

6.1* Demonstrate that the best-fitting (i.e., E-minimizing) w_j parameters for equation (6.1) are given by equation (6.2).

6.2* Suppose that all independent variables X_j and the dependent variable Y are mean-centered (i.e., they have a mean of zero), and all X_j are uncorrelated. Show that equation (6.2) then simplifies to the more intuitive $w_j = \mathrm{Corr}(X_j, Y)\,\mathrm{Std}(Y)/\mathrm{Std}(X_j)$. Here, Corr(.,.) denotes correlation function and Std(.) the standard deviation.

Parameter Estimation by the Maximum Likelihood Method

Consider a coin that generates heads or tails on each toss (as, indeed, an ordinary coin does). Our (very minimal) model entails that heads occurs with probability p, independent of each trial. We now attempt to estimate the single parameter p of the model. Suppose that we observe heads, heads, tails, heads. The probability of this sequence of events, given the parameter p, equals

$$\Pr(Data \mid p) = p \times p \times (1-p) \times p.$$

By the expression $\Pr(A \mid B)$, we mean "the probability that event A happens, given that we know B." So here, we look at the probability of the data, given that we know the

Estimating Parameters in Computational Models 93

probability p of the coin landing heads up. There is a slight ambiguity by what we mean exactly by *Data*; we can mean "this exact sequence of heads and tails," and in that case, the equation is correct. Or we could mean "3 heads and 1 tails in the sequence," in which case the equation should be multiplied by a constant (here, the number of different combinations that lead to 3 heads and 1 tails in a sequence). However, because a constant merely shifts the probability function up or down, it cannot influence our results, and we can safely ignore it.

More generally, if we observe n_H heads and n_T tails, the probability of the sequence of events equals

$$\Pr(Data \mid p) = p^{n_H}(1-p)^{n_T}.$$

Conventionally, $\Pr(Data \mid p)$ is considered as a function of *Data*, with the parameters (here, just p) fixed.

We now introduce the key concept of the *likelihood* function. The likelihood has the same value as the probability function, but it considers $\Pr(Data \mid p)$ as a function of the parameters, with *Data* fixed. Effectively, it says how likely the current parameter value (p) is, given the data that were observed. This likelihood function is denoted by L; because it takes on the same function value as the probability, it is simply

$$L = p^{n_H}(1-p)^{n_T}.$$

Again, note that L is a function of p, with *Data* fixed, but I abbreviate it to L for short.

With the concept of the likelihood function at hand, we now can ask ourselves the following question: what parameter value (here, p) has the highest likelihood for this data? Or, stated otherwise, what parameter value is the most likely to have generated our data? To find that parameter value, we need to optimize the likelihood function. Note that now we seek the maximum of the likelihood function L (rather than the minimum of a function, as in earlier chapters) because we need the parameter value with the highest likelihood.

In practice, before proceeding with optimization, one first takes the logarithm of the likelihood function. Logarithms turn products into sums, and taking the derivative of a sum is much more convenient than taking the derivative of a product. Importantly, a logarithmic transformation keeps the optimum at the same parameter value (here, p; see figure 6.2a). The logarithm of the likelihood is called the *log–likelihood*. The log-likelihood function to be optimized is eventually

$$\log L = n_H \log p + n_T \log(1-p). \tag{6.3}$$

As mentioned in chapter 1, there are three methods of optimizing a function: graphically, analytically, and computationally. I will consider each of the three approaches in the current context.

Graphically optimizing a function doesn't scale well to a model with many parameters (i.e., with higher dimensionality of the parameter space), of course, but let's first look at it graphically to test our intuition. Figure 6.2b (left panel) shows the log-likelihood function when there were $n_H = 7$ heads and $n_T = 3$ tails. As can be seen, the

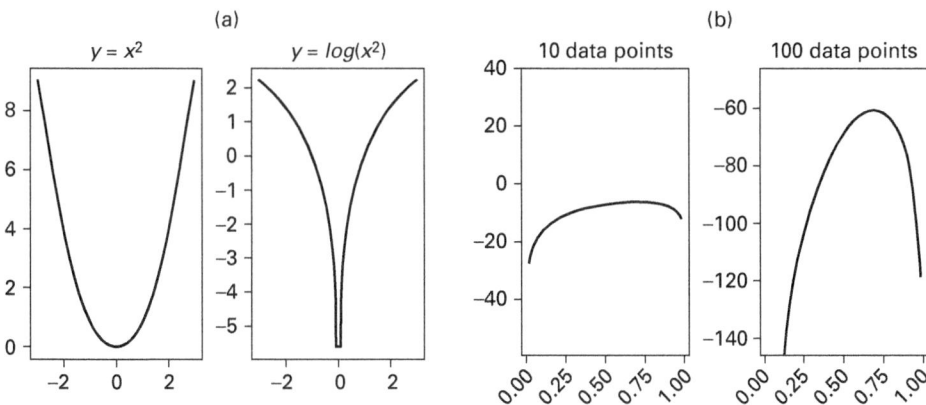

Figure 6.2
(a) Taking the logarithm of a function does not change the optimal (here, the minimum) point of that function. (b) Log-likelihood functions for the coin-tossing example.

optimal value p (probability of heads) is around .7. Does that make sense to you, given the observed data (n_H and n_T)? Figure 6.2b (right panel) shows the log-likelihood function in case there were $n_H = 70$ heads and $n_T = 30$ tails. The best-fitting p is again around .7; but the function is much more peaked, meaning that more data are available for estimating the parameter. More formally, it can be shown that the curvature (i.e., the second derivative) of the log-likelihood function determines the standard error (and thus the precision) of the parameter estimates (see the following discussion and Mood et al., 1973). This is what I meant when I said that the statistical properties of the maximum likelihood estimator are better understood.

Next, let's consider the log-likelihood function analytically. The likelihood function is so simple in this particular case that we can work out the optimal value p in an analytical manner. Setting the derivative of the log-likelihood to zero and solving for p, we find

$$p = \frac{n_H}{n_H + n_T}.$$

(6.4)

This result seems plausible: if there are relatively more heads (i.e., n_H) relative to tails (i.e., n_T), the estimated probability of the coin turning up heads increases.

We can also investigate the log-likelihood function computationally. In fact, for models of some complexity, parameters *must* be estimated computationally rather than graphically or analytically. Again, a popular algorithm to do this is the gradient ascent. A disadvantage of using gradient ascent (here, as well as in other contexts) is that it requires "smooth" gradients. If functions increase and decrease rapidly, or even discontinuously, the gradient ascent algorithm doesn't work very well (or even at all). This is because the algorithm typically needs to approximate such gradients, and such approximation is harder if the gradients are very steep. Importantly, however, recall that in this chapter, we are not considering the optimization process as something that

Estimating Parameters in Computational Models

the subject is doing. This provides us much more freedom in the optimization process; we can turn to other algorithms besides gradient ascent. For example, one popular approach for finding the maximum-likelihood parameters is grid search. Here, one divides the parameter space in a grid of points and loops over all the points of the grid. The parameter value with the highest log-likelihood is retained and is considered the best (maximum-likelihood) estimate. Grid search is not an option when the optimization is considered as a model of cognition (as in chapters 2–5); one cannot plausibly assume that a biological agent systematically searches all parameter settings for the optimal one. But when we consider optimization as simply a method to find the best (i.e., empirically most appropriate) parameters, this is no problem at all.

An advantage of grid search is that it cannot get stuck in a local maximum. Indeed, the idea is to search the whole parameter space; thus, if the range of each parameter is taken to be sufficiently large, the global maximum will be found. Unfortunately, the approach also has a number of downsides. The most severe one is probably the curse of dimensionality mentioned previously. However, if the number of parameters is low (say, $D = 2$), and if one makes the grid sufficiently fine-grained, then this method works very well. A recent, more sophisticated approach to grid search was proposed by Mestdagh et al. (2019). For a number of cognitive models, these authors calculated the data that would be predicted for a large number of grid points (e.g., parameter values; parameter-to-data function). Based on the actually observed data, one can then invert the parameter-to-data function and proceed from the actually observed data to the best-fitting parameters for that specific model. This approach is much faster than standard grid search; however, it is applicable only for models where the parameter-to-data function has been tabulated.

Another robust approach is evolutionary computation. Here, estimation time is divided into "generations"; each generation consists of a number of parameters, each with a different fit (likelihood). The best parameters (highest fit, highest likelihood) can leave offspring in the next generation with some variation. This process continues until offspring do not improve on their parents for a number of generations (Mullen et al., 2009). This algorithm combines the advantages of gradient ascent and grid search. Like grid search, it can search the whole parameter space, thus reducing the risk of getting stuck in a local maximum. Also like grid search, it doesn't require smooth gradients because gradients are not calculated at all. However, like gradient ascent, it scales well to higher-dimensional models because there is no grid that needs to be exhaustively searched. The approach therefore has increased in popularity in recent years (Calderon et al., 2017; Solway & Botvinick, 2015).

A main advantage of grid search and evolutionary computation over gradient ascent is that they require computation of the likelihood, not its gradient. A recent class of algorithms, called *approximate Bayesian computation*, does not even require computation of the likelihood; it only requires that one can sample data from the model. For example, sampling from a neural network model is relatively easy; one presents an arbitrary input pattern and observes the outcome. The interested reader can consult

96   Chapter 6

Turner and Van Zandt (2018) or Farrell and Lewandowsky (2018) for an introduction to this algorithm. Although this approach has been relatively underexplored, it may be very relevant for fitting parameters of complex models, such as neural network models, where even writing the correct likelihood function may be challenging. Thus, this and similar approaches are likely to offer exciting possibilities in future neural network research.

As alternatives to gradient ascent, we have discussed alternative methods up to now that require *less* information about the function (e.g., only the function value, but not the function gradient). One could instead consider methods that use *more* information about the function. In particular, whereas gradient ascent uses only first-order derivative information, other methods (such as the Newton-Raphson estimation algorithm) also use second-order derivative information. Computing such second-order derivatives is hard, but once obtained, they can allow very efficient stepping in the parameter space toward the optimal point. We refer readers to specialized engineering (Gill et al., 1982) or cognitive modeling literature (Farrell & Lewandowsky, 2018), or to the documentation of the scipy package in Python.

Exercises

6.3 Show that optimizing equation (6.3) leads to equation (6.4).

6.4 Find conditions under which minimizing the error function [equation (6.1)] can be considered an instance of maximum likelihood estimation.

6.5 Demonstrate that the sample mean is the maximum likelihood estimate of the population mean when errors are independent and identically sampled from a Gaussian distribution. Can you make the same claim for any distribution from which the data were sampled?

Parameter Identifiability and Estimation Precision

An important issue in parameter estimation is identifiability. To explain this, we will need to switch to a slightly more complicated model than the 1-parameter coin toss model. Consider the four-armed bandits shown in figure 6.3. A subject can choose any of the four options (called "arms" in a casino context, hence its name). The subject can repeatedly choose between each of the $N = 4$ arms (in general, there are N arms,

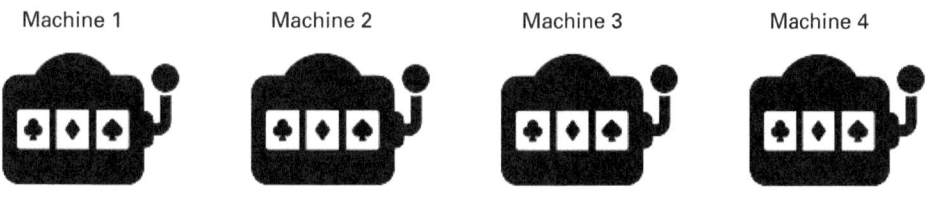

Payoff probabilities:

$p_1 = .8$ $\qquad\qquad$ $p_2 = .2$ $\qquad\qquad$ $p_3 = .4$ $\qquad\qquad$ $p_4 = .6$

Figure 6.3

The four-armed bandit, here depicted as four one-armed bandits, each with its own payoff probability p_i.

Estimating Parameters in Computational Models 97

in which case we'll refer to the collection of machines as an "N-armed bandit"[1]). How does the subject choose?

To address this question, a very popular decision-making model assumes that subjects associate a weight w_i (corresponding to the ith bandit preference) to each of the bandits. On each trial, one of the N bandits is chosen as follows:

$$\Pr(\text{Choose}\, i) = \frac{\exp(\gamma w_i)}{\sum_j \exp(\gamma w_j)}. \tag{6.5}$$

This implies that the bandit with the highest preference (w_i) has the highest probability of being chosen. Just as in chapter 4, γ can be considered as a slope parameter; lower values of γ make the process of choosing among the different bandits noisier. This parameter is also sometimes called the *inverse temperature*. The name of this parameter derives from the fact that when $\gamma = 0$ (maximal temperature), all responses are random, corresponding to a "hot" environment (with forceful "movement"). Instead, when $\gamma \to +\infty$, responding becomes deterministic, corresponding to a "cool" environment (with little "movement"). For a more precise analogy with the physical notion of temperature, see Verdonck and Tuerlinckx (2014).

To illustrate these concepts, table 6.1 shows the choice probabilities [equation (6.5)] when there are four bandits, each with a preference w_i. The first row depicts probabilities with a low γ ($\gamma = 0.2$), the second row with a high γ ($\gamma = 2$). In each case, the bandit with higher w is preferentially chosen, but sometimes another bandit is chosen. For this reason, this is sometimes also called a *softmax rule*: one takes the maximal element, but in a soft (noisy) manner. Competitors are not totally suppressed. With higher γ ($\gamma = 2$), the process is less noisy, and losers are suppressed more strongly.

We could now consider estimating the parameters γ and w_i ($i = 1, \ldots, N$) simultaneously. However, this would not be possible because no unique solution maximizes the likelihood. In fact, in this case, there is an infinite number of parameter combinations that maximizes the likelihood. To see why this is true, consider that $C\gamma$ and w_i/C for each $C \in \mathbb{R}_0$ lead to exactly the same likelihood values. To see why, plug them into the probability [equation (6.5)] and observe what happens. Thus, it makes no sense to say that one parameter combination is best. Formally, this is the same reason as why we could not optimize the slope and weights parameters simultaneously in chapters 4 and 5.

Table 6.1
Illustration of the effect of γ on choice probabilities in the softmax rule ($N = 4$)

w	2	.2	0.5	1
$\gamma = 0.2$				
$\Pr(\text{Choose } i)$	.31	.21	.23	.25
$\gamma = 2$				
$\Pr(\text{Choose } i)$	.83	.02	.04	.11

If a parameter is not identifiable, the researcher must place some constraints on the estimation process. This can be done by fixing one of the parameters, such as $\gamma = 1$. But even that is not enough in this case; any shift $w_i + C$ for each $C \in \mathbb{R}$ would still lead to the same probabilities [equation (6.5)]. When one of the w_i values is also fixed to some value, such as $w_1 = 1$, the problem is solved and the model is identifiable. It is sometimes easy to find that parameters are not identifiable, but typically hard to prove that they *are* identifiable.

In this case, the identifiability problem is rather obvious, but often the problem is less clear. Even harder to detect is when parameters are not unidentifiable, but are *weakly* identifiable. The parameter estimation problem in figure 6.2b (left panel) can be considered weakly identifiable; many parameter values lead to approximately the same likelihood. In this way, it becomes rather arbitrary to claim which parameter combination is best. Moreover, just slightly changing the data can lead to very different estimates. This can again be seen in figure 6.2b (left panel). Adding some random changes to the data in figure 6.2b (left panel) would gently move the likelihood function, but it can lead to a dramatically different optimum. In contrast, gently moving the strongly identified likelihood curve of figure 6.2b (right panel) would still lead to an estimate around the same value. In that sense, the parameter in figure 6.2b (right panel) is strongly identified.

Parameter estimation and identifiability are technical issues about which quite some theory and methods are available, but that discussion is beyond the scope of the current book. We refer readers who want to know more or apply these methods to other sources (Daw, 2009; Farrell & Lewandowsky, 2018; Mood et al., 1973).

In general, an easy and convincing way to check whether parameters of your model are identified is via a parameter recovery simulation. In this approach, you first take a specific point in the parameter space of the model (say, θ, of dimension K). Then you simulate a data set (of size N) based on this parameter. For example, in the coin-tossing case, simulating a data point simply consists of generating heads with probability p (and thus tails with probability $1 - p$). A data set can be simulated in this example by repeating this process N times (i.e., for generating N data points). Then estimate the original parameter based on the simulated data (yielding estimate $\hat{\theta}$). One repeats this procedure a number of times (say, M times; M could be set to 1,000), for different points in the parameter space. Finally, calculate the correlation $r(\theta_i, \hat{\theta}_i)$ for each parameter i separately (hence, K correlations, each of which is based on M data points). If the data sets are sufficiently large (i.e., N is large), each of the K correlations should approximate 1. The parameters of the model are then well recovered by estimation. Even if you have no doubts about parameter identifiability, if you want to estimate and interpret parameters from a model, a parameter recovery study is usually the step to take before estimating these parameters on empirical data. It tells you how much data (N) is needed to reliably estimate your parameters.

The notion of parameter estimation reliability is captured more formally by the concept of a standard error. The standard error is the standard deviation of the parameter estimate distribution. In other words, if we sampled all possible data sets of size N, estimated the parameter for each possible data set, and drew the histogram of the resulting parameter estimates, the standard deviation of that distribution would be the standard error (for sample size N). I already noted that the standard error can be explicitly calculated in a maximum likelihood framework based on the curvature (second derivative, or peakedness, of the likelihood function). Again, see figure 6.2b, where the function in the left panel has low curvature (low peakedness, high standard error) and the right panel has high curvature (high peakedness, low standard error). It is a remarkable fact of statistics that with just a single sample (i.e., our data set), one can obtain an estimate of what happens if we would have collected all the possible data sets.

However, as usual, this neat result comes with a price. We have to be able to calculate the likelihood and its second derivative.[2] If this assumption is not valid (or computation of the derivatives not feasible), a bootstrap approach can be used to estimate the standard error (Efron & Tibshirani, 1993). Here, a novel data set is created by sampling (*with* replacement) from the original data, and a novel parameter estimate is computed. This procedure is repeated a number of times (say, 10,000), and the resulting parameter estimate distribution can be used to calculate various statistics of interest, including the standard error of the parameter (thus, the standard deviation of the estimate distribution).

Comparing the parameter space exploration approach (in the first part of this chapter) and parameter estimation approach (in the second part of this chapter) to dealing with model parameters, one can see that in some sense, they have opposite goals. In the parameter space exploration approach, one attempts to demonstrate that the exact parameter value is irrelevant. Hence, in this approach, it is considered good if different parameter combinations lead to the same likelihood (or other goal function that one attempts to optimize). In contrast, in the parameter estimation approach, one wants exactly the opposite—namely, a likelihood value that is as different as possible for parameter values. Which approach is better depends on exactly what goal the modeler wants to achieve with the model. In the next section, I discuss a few applications where parameters are estimated and interpreted from a substantive perspective.

Applications

Working Memory: Slots, Resources, or Interference?
An early theoretical view holds that working memory consists of a few slots where information can be stored. If the information exceeds the slot size, information is lost. The number of slots was originally estimated to be around seven (Miller, 1956), but more recently is estimated to be more realistically around four (Cowan, 2001). An interesting test of the concept of working memory slots was carried out by Zhang and Luck (2008). Following a procedure from Wilken and Ma (2004), they asked subjects to retain color information at three locations (i.e., three color patches were presented at different spatial locations) in working memory for a brief time. After the retention interval, subjects had to reproduce one of the colors on a color wheel afterward. Based

on these data, they estimated parameters of a model containing two parameters: the probability (p) that each item would be stored in a slot; and the standard deviation (sd) of the representation (inverse noise) of the slots. For example, if there were exactly three slots, each would have standard deviation sd; if six items were presented, the probability for each of the items to enter in one of the slots would be ½.

Together, these parameters can be used to fit a computational model (the slots model) on the behavioral data of a group of subjects who took this working memory task. A critical prediction of this slots model is that the probability p of each item entering a slot gradually decreases if more items are presented in the initial item display. In contrast, the standard deviation sd should remain constant, regardless of the number of items presented. After fitting the model to the data, the authors indeed observed that p gradually decreases with larger set size. For sd, they observed that it increases up to (item) set size 3, but remains constant for set sizes of larger than 3. The initial increase was explained by the authors in their modified "slots + averaging" model; they assumed that when fewer than 3 items were presented, then more than one slot could be assigned to specific items, hence functionally decreasing sd in that case.

An alternative *resource* model was postulated around the same time by Bays and Husain (2008) and Ma et al. (2014). They proposed that instead of a collection of slots, working memory can be considered as a resource to be distributed among any of the presented items. In the experiments reported by these authors, subjects were also required to keep a visual display of colored patches in working memory. But here, subjects reported the horizontal displacement (i.e., displacement is left or right) of specific items after a short retention interval. Based on those binary reports, the authors could estimate the representational precision of Gaussian curves (inversely related to their standard deviation), centered on each remembered item, for each subject separately. In this analysis, smaller standard deviations correspond to more available resource. It was observed that with set sizes 1 to 12, precision of the Gaussians decreased gradually, with no evidence of a nonlinearity around set size 3 or 4. This result is the opposite of what the slot model would predict, which does predict a nonlinearity at around 3 or 4 items.

Finally, an *interference* approach to working memory assumes no slots or resource to be distributed across various memoranda. Instead, it postulates that information in working memory is impaired by interference from other, partially overlapping items in working memory (Oberauer & Lin, 2017). Note that this conceptualization is more related (than either the slots or resource models) to the neural network models from earlier chapters (e.g., discussion of interference in chapter 3). In particular, the authors assumed that items in the Zhang and Luck (2008) and Bays and Husain (2008) experimental paradigms would gradually become associated with the context, which in this case would be their stimulus location.

This process of gradually associating items with context could be implemented via Hebbian learning (discussed in chapter 3), although the authors did not explicitly model the learning process. Hence, in their neural network model, context (stimulus location) constituted the input layer, and the items to be remembered make up the output layer of this two-layer model. Items at nearby locations would thus interfere with each other more, leading to errors during recall. They added some extra assumptions: (1) items are partially activated regardless of location; (2) there is background noise; and (3) one item is in the focus of attention (for details, see Oberauer & Lin, 2017). These authors estimated the parameters of all three models discussed up to now (slots + averaging, resource, and interference). The crucial question, of course, is which model fits their data best; the answer will be discussed in chapter 7.

Estimating the Diffusion Model

Several algorithms have been proposed to estimate parameters of the diffusion model, either starting from an error minimization or maximum-likelihood perspective (Ratcliff & Tuerlinckx, 2002). The simplest algorithms estimate three parameters for each subject and each condition separately: drift rate, threshold, and non-decision time (Wagenmakers et al., 2007; see also discussion in chapter 2). However, more powerful algorithms allow great flexibility in estimating parameters of the diffusion model based on empirical data. For example, in addition to drift rate, threshold, and non-decision time, the starting point can be estimated. This allows for modeling a bias toward one of the two responses; an example of this approach was already discussed extensively in chapter 2 (Mulder et al., 2012). Moreover, one can incorporate trial-to-trial variability in each of these parameters, making the model more flexible to fit data.

One goal of model parameter estimation could be evaluating to what extent each parameter varies across conditions or groups of subjects. As an example of this approach, Ratcliff et al. (2004) observed that older subjects are slower and more accurate on a psycholinguistic task (i.e., a lexical decision task). However, is this because they are more cautious, or do they also have an impaired ability to perform such tasks? Estimating the parameters of the diffusion model based on their data allowed for disentangling these possibilities. The authors demonstrated that older subjects have a slower non-decision time and a higher threshold; but the drift rates were very similar across populations. The authors concluded that older subjects are generally slower and more cautious but have similar psycholinguistic ability.

Some statistical diffusion model estimation packages allow the inclusion of covariates that explain parameter variation across trials, as well as adding restrictions across conditions and across subjects (Vandekerckhove & Tuerlinckx, 2008; Wiecki et al., 2013). For example, a plausible restriction could be that for each subject, there is a fixed threshold and non-decision time (i.e., shared across conditions), but drift rate is different for different conditions. Moreover, in the hierarchical estimation approach of Wiecki et al. (2013), all parameters of the same type (e.g., all drift rates) can be modeled as being sampled from a common distribution (which in statistics is called a *random effects model*). The parameters of the common distribution (which are sometimes called *hyperparameters*) can then (also) be estimated. Because data are pooled across subjects in this approach, estimation of the population-level characteristics (e.g., mean drift rate across subjects) can be estimated with higher precision than at the single-subject level.

Such pooling of parameters can be done in a principled way in a Bayesian (in contrast to maximum likelihood) framework toward statistical estimation. I will discuss Bayesian computational models in chapter 11, but the topic of Bayesian statistics (and its potential for pooling of parameters) is outside the scope of the current book; I refer to Kruschke (2015) for more information on that topic.

It is useful at this point to mention another interesting recent approach to parameter estimation, consisting of the joint modeling of different data modalities, all constrained by the same underlying model. For example, Turner et al. (2013) proposed a joint model for choices and response times (RTs) in a random-dot motion task, with white matter tract strength data, as measured with diffusion tensor imaging (DTI) in the brain of the same subjects who also performed the random-dot motion task. For the behavioral data, they used the linear ballistic accumulator (LBA) model, a simpler variant of the diffusion model (Brown & Heathcote, 2008); for the tract data, they used a Gaussian distribution. They stipulated in the joint model that parameters of the underlying LBA and Gaussian (tract) models could be correlated. Among others, they observed across subjects (in two of three conditions) a negative correlation between LBA response threshold and tract strength. Subjects with a lower LBA threshold tended to have a stronger tract strength. Statistically, to the extent that the two data modalities have a common structure, this joint modeling approach tends to yield more precise estimates (i.e., with lower standard error), simply because there are more data on which to base the estimates.

Learning Models of Decision Making

Until now, I have described the statistical estimation approach on static models (i.e., where parameters remain stable across trials): namely, a coin-tossing model, models of working memory, and the diffusion model. In chapters 3–5, we worked instead with dynamic models, in the sense that they learn optimal weights w_{ij} across trials. These parameters w_{ij} cannot be estimated; they are assumed to be learned by the model or participant as the task unfolds. However, one could consider whether the fixed parameters from the dynamic models, like learning rate (β) and slope (γ), can be estimated using the tools that we have described in this chapter.

Unfortunately, statistical parameter estimation in dynamic learning models is typically more challenging than in static (i.e., non-learning) models. The likelihoods of dynamic models are often not very identifiable. Until recently, parameters like learning rate and slope were often just fixed at some value, and it was demonstrated (or at least hoped) that the specific values do not matter very much.

However, in part due to powerful computers and software that allow estimating more complicated models and methods for testing identifiability, recent years have seen an increasing practice of estimating parameters in learning models. In fact, estimation of parameters in learning models is a major theme in recent studies of decision making (Behrens et al., 2007); and in how decision making is disturbed in clinical populations such as attention deficit hyperactivity disorder (ADHD) (Hauser et al., 2014).

Estimating Parameters in Computational Models

Further, a whole new field called *computational psychiatry* investigates, using computational models, how learning and decision making are impaired in clinical conditions (Maia & Frank, 2011).

To illustrate this research line, consider again the N-armed bandit from figure 6.3. Now, the subjects do not have preferences w_i that we attempt to estimate; instead, the N arms deliver rewards with payoff probabilities $p_1, \ldots, p_N$ and subjects must learn which arm is best (and should thus be sampled from most often).

To choose the best (i.e., the highest paying) bandit, a model can learn the value w_i of each bandit based on trial-to-trial feedback. To do so, for each trial, it applies the following delta rule:

$$w_i(n) = w_i(n-1) + \beta x_i \big(R(n) - w_i(n-1) \big),$$

where, $x_i = 1$ if the bandit was chosen on trial n, and $x_i = 0$ otherwise. The variable $R(n)$ indicates whether a reward was provided ($R = 1$) or not ($R = 0$) on trial n. This model also contains the learning rate β as an estimable parameter. One final parameter is the starting value w_0 of the weights. However, if the learning rate and number of trials are sufficiently high, then w_0 hardly influences the data, and it is usually just fixed at zero or at a random value. Finally, we again choose a bandit i depending on its value, but in a probabilistic manner:

$$\Pr(\text{Choose}\, i) = \frac{\exp(\gamma w_i)}{\sum\limits_j \exp(\gamma w_j)}.$$

Two parameters, learning rate β and slope γ, can now be estimated by either error minimization or maximum likelihood maximization. In some cases, parameters β and γ are still unidentified or weakly identified (Daw, 2009), so if both parameters are estimated and interpreted, the modeler must carefully consider whether such parameters can be faithfully recovered by estimation with the available sample size. Consider figure 6.4:[3] Here, three data sets were generated with 100, 1,000, or 3,000 trials according to the four-armed bandit model. Contour plots (lines of equal likelihood) are drawn for each data set. The dot is the true parameter (i.e., the parameter that generated the data); the triangle represents the maximum likelihood estimate. The contour range is kept identical in the three plots; hence, fewer contour lines are visible with flatter likelihood curves (which here correspond to lower numbers of trials). Numbers on the contours indicate the negative log-likelihood values (–log-likelihood—hence, lower is better). These plots suggest that simultaneous identification of learning rate and temperature is feasible if one has 1,000 or more trials, but it may be tricky with lower numbers.

To see this effect more systematically, consider table 6.2. Here, as an alternative approach to calculating a correlation between real and estimated parameters to investigate parameter recovery (as proposed earlier in this chapter), I took one parameter value and evaluated the average estimate and its standard deviation. Model estimates (across $M = 50$ replications) are shown for data sets with 100, 1,000, and 3,000 trials, respectively. The estimates appear to be unbiased: The average estimate in each case is

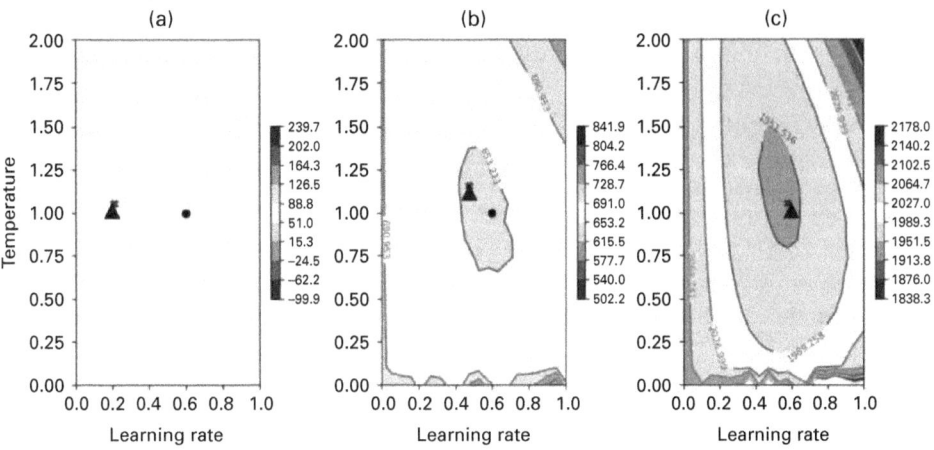

Figure 6.4
Contour plot for the estimation of learning rate and temperature in the four-armed bandit case.
(a) 100 trials; (b) 1,000 trials; (c) 3,000 trials.

Table 6.2
Model parameter estimates for learning rate (true parameter = 0.6) and temperature (true parameter = 1)

Number of Trials	Learning Rate Mean	Learning Rate Standard Error	Temperature Mean	Temperature Standard Error
100	0.69	0.40	1.05	0.52
1,000	0.60	0.13	1.01	0.16
3,000	0.63	0.07	1.00	0.08

close to the actual parameter value. However, the standard error (which can be approximated by the standard deviation across the 50 replications) is quite large for the smallest sample size (100 data points).

Similar estimation results were obtained when I simulated a reversal learning experiment in which the probabilities switch on some trials. Readers can further explore goodness of parameter recovery using the code available on GitHub. Indeed, the specific trial numbers in this particular case are not what matters: what matters is that you can (and should) check the quality of parameter recovery in your own model of interest.

A general rule is that the more complex (i.e., rich) the model is whose parameters we intend to estimate (e.g., measured in number of parameters), the richer the data have to be in order to estimate those parameters. Other than simply collecting more trials per subject, another option to make the data set richer (and thus obtain more stable parameter estimates), is to combine different dependent variables, such as choice data and RT data for estimating common parameters (Ballard & McClure, 2019). This approach is similar to the joint modeling across data modalities approach mentioned earlier.

One interesting application of the learning model of decision making appears in the computational psychiatry literature. For example, using a model very similar to the one just reported, Hauser et al. (2014) estimated learning rate and temperature separately in control versus ADHD juveniles. They observed that performance in the ADHD group was noisier overall (lower γ) than in control subjects, but the ADHD group had a similar learning rate (β) relative to the control group.

To sum up, parameters in computational models can be treated in two ways: qualitatively (inspecting the parameter space) or quantitatively (estimating the best parameter values based on data). I discussed one statistically well founded approach to parameter estimation: the maximum likelihood approach. When estimating parameters, we must be careful that our models are not too rich (complex), relative to the amount of data we have available to estimate its parameters. To evaluate this, we must investigate (via a parameter recovery study) whether we can accurately (i.e., with small standard error) estimate the parameters from the models that we fit. One recent approach to make the data richer, of course, is to collect more data; another recent approach is joint modeling, where various data modalities are combined to simultaneously constrain model parameters.

7 Testing and Comparing Computational Models

Prediction is very difficult, especially about the future.
—Attributed to various sources

Chapters 1–5 covered choosing an appropriate model for your data. Chapter 6 discusses how to choose appropriate values for the parameters in these models. The next question we ask is whether the model that we chose is appropriate for the data, a process called *model evaluation* (which is a part of statistical model analysis). Is it appropriate to use the model architecture and model dynamics to make inferences about how humans (or other biological agents) behave? Recent statistical methods combine model estimation (chapter 6) and model evaluation (chapter 7) (Piray et al., 2019), but this approach goes beyond the scope of this chapter, and we will keep the two model analysis aspects separate.

Within model evaluation, two more specific questions can be distinguished. First, does the model, in an absolute sense, provide a good fit to the data? This aspect is considered in model testing. Second, does the model, in a relative sense, provide a better fit than other models under consideration? This is studied in model comparison; we discuss these topics consecutively here.

To start, however, we note that model testing and model comparison are exactly the steps that an empirical researcher or applied statistician would perform after parameters from a model are estimated (even though estimated parameters are not required for all types of model evaluation; see below). Consider again, as presented in chapter 6, the familiar linear model from statistics:

$$Y(n) = \sum_{j=1}^{J} w_j X_j(n).$$

Researchers applying linear regression might consider, for example, the question "Does a model containing regressors X_1 and X_2 provide a good fit to the data (i.e., the dependent variable)?" This is a model testing question. Alternatively, they could consider the question "Is a model containing both regressors X_1 and X_2 better than a model containing just regressor X_1?" This is a model comparison question. The relative fit (model comparison) of these two models then provides information about the importance of adding regressor X_2 to the regression equation. Thus, regression (including analysis-of-variance) models employ the model estimation and evaluation tools considered in chapters 6 and 7 (Maxwell et al., 2004). This is why we treat them relatively briefly here, even though these topics are important. Model estimation and evaluation are tools used throughout the statistical sciences, and several handbooks describe them in much more detail than this volume (Farrell & Lewandowsky, 2018; Maxwell et al., 2004).

From a different angle, the linear models that are common in data-analysis approaches can be considered as simple computational models, in which linear combinations of variables X_j combine to generate the dependent variable Y. If it turns out that a linear model is an accurate representation of some cognitive process, then all the better: we can then use the very extensive theory and software available for linear models. Of course, the linear model from data analysis is itself usually not of interest. Here, the model is typically just a statistical vehicle to make meaningful statements about the relations between different factors. For example, in a survey study, one is not really interested whether education level and socioeconomic status linearly combine to determine salary. Instead, one is interested in whether socioeconomic status has an effect on salary (a statistical main effect question) and to what extent that effect depends on education level (a statistical interaction question). The linear model is only needed to state and answer such questions precisely. Instead, in computational modeling, the model will incorporate our cognitive theory, which is not necessarily linear (although it typically contains a linear core; see the discussions about the models in chapters 2–5). Thus, we replace the off-the-shelf linear models with some (hopefully) more appropriate models for the data that we want to understand. In the current era of open software, more and more packages for fitting increasingly complex computational models are becoming available (Wiecki et al., 2013). In some cases, however, we will have to get our hands dirty and implement the statistical routines ourselves.

Model Testing

Omnibus Tests

Omnibus tests consider whether the model as a whole provides a good fit to the data. Consider again the diffusion model. After estimating its parameters, we can simulate response time (RT) distributions under the model for both correct and error trials (or derive the RT distributions analytically). One can then qualitatively investigate whether the observed distribution is close to the predicted one. Such a qualitative approach will often already provide a good indication of whether the model fits the data. Consider figure 7.1, which shows observed RT distributions for two responses for two subjects (remember that each stimulus has two possible responses in the diffusion model; the responses depicted are not to the same stimulus, and therefore the probabilities do not

Testing and Comparing Computational Models   109

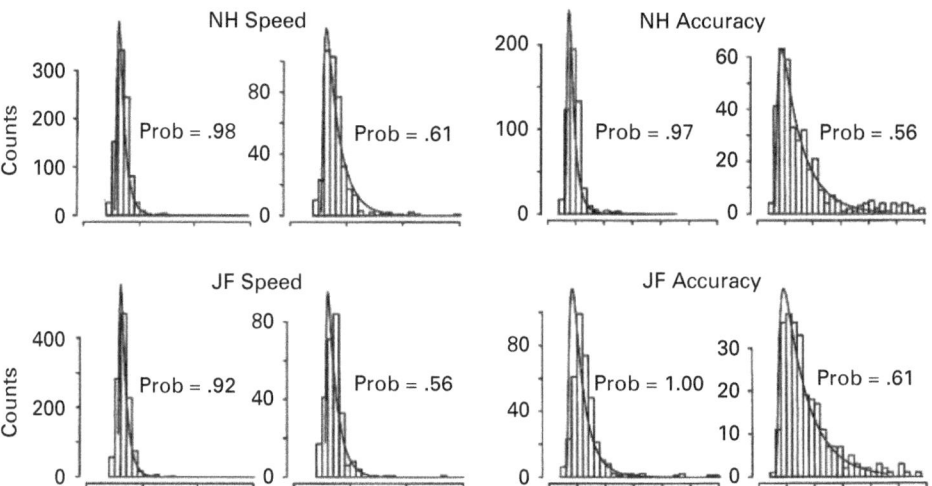

Figure 7.1

Empirical RT histograms for two subjects (NH and JF), in two conditions, for two separate responses (probability of giving each response is shown next to each distribution). Reproduced with permission from Ratcliff and Rouder (1998).

add to 1). These empirical distributions can be compared with the theoretical predictions from the diffusion model (full lines). Their close correspondence suggests that the model can faithfully describe the RT data in this case.

To quantify this difference in a statistic (an omnibus test), one could arrange the RT data into quantiles. For example, one could calculate the deciles of the RT distribution, where each decile contains exactly 10% of the RT distribution (e.g., quantile 1 contains all RTs between 100 and 150 ms, quantile 2 has all RTs between 150 and 170 ms, and so on) (Ratcliff & Tuerlinckx, 2002). Then calculate the percentage of simulated data in each quantile, and add the squared differences (across all the quantiles) between observed and predicted percentages in each quantile. A value of zero for this statistic calculated on the simulated data would indicate perfect correspondence between data and simulation. Larger values indicate misfit; the simulated RT data then do not have their probability mass where the observed RT distribution has its probability mass. For statistical testing, one would also like to know the distribution of such statistics. I will discuss how to construct these distributions later in this chapter.

A drawback of an omnibus test is that it doesn't provide very specific information. Suppose that the statistic described here yields a high value, and statistical testing rejects the tested model. This indicates that only *some* aspect of the model is wrong. We don't know whether it's a relevant aspect that the omnibus test was sensitive to. This problem is addressed with the more specific model assumption tests described in the next section.

Model Assumption Tests

I will first explain testing model assumptions in the context of the linear model, where the dependent variable is modeled as a linear combination of the independent variables (the regressors X_j). One key assumption in the standard linear model is that the errors (deviations of the observations from the theoretical mean, which in some sense is also a prediction error), follow a normal distribution. Moreover, they are assumed to be independently distributed. For example, it may be the case that the error on a trial is larger when the error on the previous trial was also larger; this would constitute a violation of the assumption.

Although errors as defined here cannot be computed (as we do not know the theoretical mean), we can compute the residuals (deviations of the observations from the model estimate; in the simplest case, the model estimate is just the sample mean). We can check the independence assumption in the residuals. Figure 7.2a shows residuals from a model where errors are independently and normally distributed; figure 7.2b from a model with some dependency (residuals seem to drift initially downward and then upward); and figure 7.2c from a model with strong dependency.

Besides independently and normally distributed residuals, another assumption of the linear model is that regressors contribute to the dependent variable in an additive manner. In other words, the outcome at each measurement unit i is just a linear combination of all regressors (on that level i). In some situations, this assumption is trivial, but certainly not always. For example, in functional magnetic resonance imaging (fMRI) research, a linear model is typically used to analyze the Blood Oxygen Level Dependent (BOLD) signal. The additivity of regressors is an important and nontrivial assumption in this case. Indeed, it may very well be that the response in a brain area at time t depends on whether there were a response in that same brain area (perhaps

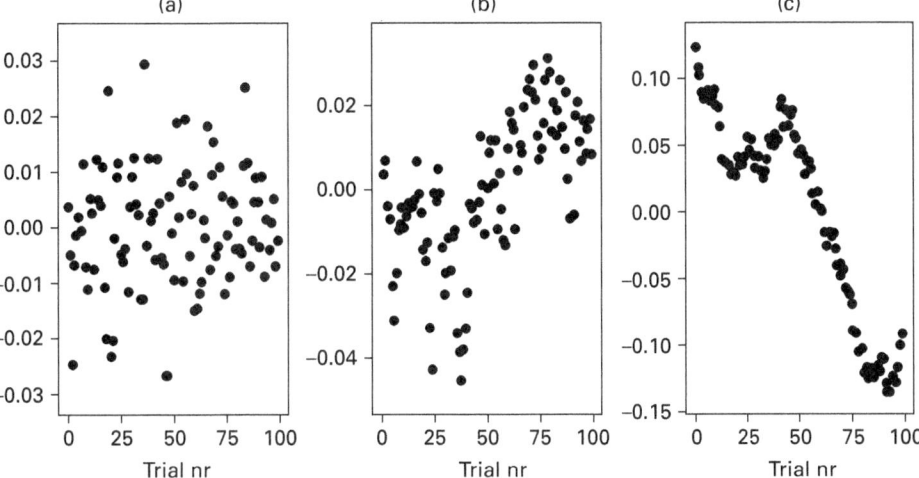

Figure 7.2
Model residuals (the mean per plot is zero by definition). (a) No dependency—each data point is sampled from an independent normal distribution. (b) Dependency—each data point is 0.8 times the previous one plus random noise. (c) Strong dependency—each data point is the previous one plus random noise. Note that the y-axis scale is different for the three panels.

Testing and Comparing Computational Models 111

due to another factor, captured by another regressor) that was active just before (at time $t - \Delta$). For example, if a group of neurons has just responded to a stimulus, the same neurons may respond differently (e.g., more weakly due to mental fatigue) when a similar stimulus is presented afterwards (Grill-Spector et al., 2006). If this is the case, it would constitute a violation of additivity. Fortunately, empirical investigations have demonstrated that neural responses are approximately additive in the BOLD signal, at least when relevant events are separated by at least a few seconds (Dale & Buckner, 1997).

The previous two model assumption tests (independence and additivity) are rather technical; they concern model assumptions that do not necessarily interest us but are needed for appropriate statistical inference. Let's now consider some more substantive model assumption tests. Ideally, a model assumption test should depend only on the model architecture, not on the specific parameter values. A nice example of this approach appears in Busemeyer et al. (1993b), which demonstrated that, independent of the parameter settings, a large class of network models (including the models discussed in chapters 4 and 5) predict that there will be no cue competition effects when cues are not actually correlated (i.e., in the stimulus set seen by a participant). Whether cue 1 has low or high validity does not influence the effect that cue 2 has on the criterion. For example, suppose that we construct a model of social cognition in an attempt to understand how human participants infer intelligence (dependent variable) from "wearing glasses" (cue 1) and from "social status" (cue 2). If the two cues are not correlated in the experimental design, the validity of wearing glasses should not influence the effect that participants ascribe to social status when judging their fellow human being on how intelligent they are.

In their companion paper, Busemeyer et al. (1993a) show that in fact, cue competition effects are empirically observed when the cues are not (by experimental design) correlated. Specifically, when one of the cues (say, wearing glasses) had a high validity for predicting the dependent variable, participants underused the other cue (say, social status). This underuse of social status did not occur when the cue wearing glasses had a lower validity. This result demonstrates that, at least with respect to cue competition, such models would need to be adapted to conform to this data pattern. Such parameter-free model testing is in principle a very informative approach because it indicates specifically what is wrong (or right) in a whole class of models in one test, independent of the specific parameter settings of the model.

As another example, Palminteri et al. (2015) distinguished between absolute and relative value coding in decision making. Specifically, in their experiment, subjects had to learn that one stimulus (G75) led to a 75% reward, whereas another stimulus (G25) led to only a 25% reward. Models that use absolute value coding (e.g., the decision-making model discussed in chapter 6 in the context of the four-armed bandit) can easily learn to choose the G75 over the G25 stimulus. However, consider a loss context where the subject must choose between stimulus L25 (loss 25% of the time) and stimulus L75 (loss 75% of the time). An absolute coding model will learn to prefer L25 over L75 and generally predict the preference order G75 > G25 > L25 > L75 (i.e., G75 is preferred above G25, and so on). Instead, a relative value coding model calculates the values of individual stimuli relative to the reward (or loss) context. Here, a L25 stimulus can acquire positive value relative to the overall negative value in the loss context. Thus, the relative coding model predicts the preference order G75 > P25 > G25 > P75. As a result, the two models could be compared in a parameter-free manner. Empirical data were consistent with the latter preference order, and thus with the relative coding model. Also, model simulations confirmed that the absolute coding model could not, but the relative coding model could, predict the observed preference order (Palminteri et al., 2017). Such model assumption tests are informative because a lack of fit between model and data immediately identifies which aspect the model got wrong, and thus which aspect of our cognitive hypothesis should be changed.

Deriving a Statistical Distribution

For statistical model testing, one places the model in the null hypothesis and constructs a statistic. A statistical distribution is then derived for this statistic under the model (null hypothesis). If the resulting test statistic is unlikely to have been generated under this null hypothesis, it is interpreted as a violation of the null hypothesis, and thus a violation of the model. In contrast, if the null hypothesis is not rejected, this is considered as tentative support for the model. As a simple example, suppose again that we have a coin with probability of heads p: our simple computational model is that the coin is biased toward heads (with $p_H = 0.8$). A relevant statistic is then

$$X^2 = N \times \left(\frac{(p_H - 0.8)^2}{0.8} + \frac{(1 - p_H - 0.2)^2}{0.2} \right), \tag{7.1}$$

which computes the extent to which the observed p_H and $1 - p_H$ deviate from their model-predicted values of 0.8 and 0.2 (respectively) multiplied by the number of data points (N). Asymptotically, this statistic X^2 has a χ^2 distribution with 1 degree of freedom (Bishop et al., 2007; change the parameters in ch7_chi2.py to better understand the distribution). Of course, a computational model will usually generate a more complex prediction than just $p = 0.8$; moreover, prediction p will be based on estimated parameters.

But even with estimated parameters, we can use the X^2 statistic. In particular, under some assumptions (e.g., Bishop et al., 2007), if there are M conditions in the experimental design and K estimated parameters, the statistic X^2 has a χ^2 distribution with $M - 1 - K$ degrees of freedom. Note that in the coin-tossing example, $M = 2$ and $K = 0$, so there is just 1 degree of freedom, as postulated here. In this case, there is no room for parameter estimation; estimation of just a single parameter ($K = 1$) would reduce the degrees of freedom to zero. Thus, if a model makes predictions about a discrete variable across M cells and K model parameters can be estimated (with typically $K \ll M$), the statistic in equation (7.1) provides an omnibus (overall) model test.

The general logic of this approach will be familiar to readers from standard statistical practices. Note, though, that the approach is conceptually different from the standard

Testing and Comparing Computational Models

one in the sense that in the current context, the model that one is actually interested in is placed in the null hypothesis. Usually, instead, the hypothesis that one seeks to reject is placed in the null hypothesis (e.g., a model where some regressor of interest has a zero coefficient). In both cases, of course, they are just models that are empirically compared.

Although omnibus or specific statistical testing can be informative for model evaluation, it is important to realize that no model is actually ever correct. A model is supposed to be a useful simplification to allow us to think about the data and provide novel, sharper predictions. Therefore, whether or not a model (or one of its assumptions) is rejected will ultimately depend on the statistical power of the data that were collected. With few data, the null hypothesis will not be rejected; with many data, it will be. It is typically difficult to know whether support for a model in this approach is not just due to low power.

Statistical testing requires knowing the distribution of the relevant statistic under the null hypothesis (i.e., under the model). If we have the correct distribution, we can compare the observed value of this statistic with its theoretical distribution. A good deal of statistical research in the early twentieth century was devoted toward deriving such distributions (typically, t, F, or χ^2 distributions), and the discussion around equation (7.1) is typical for this body of work.

Although these theoretical results remain useful (and used), the advent of fast computers rendered this work less critical for statistical practices. Nowadays, the relevant distributions under the null hypothesis can often be obtained by a resampling method. We already mentioned bootstrapping in chapter 6; this is one instance of the class of resampling methods where it is used to estimate the standard error of a model parameter estimate. But resampling can also be employed for statistical model testing. Although several specific algorithms exist depending on the exact situation, the general logic in the context of model testing is as follows (Efron & Tibshirani, 1993).

1. Calculate the observed T statistic, T^{obs}.

2. Simulate new data that is consistent with the model in the null hypothesis (but not consistent with the alternative hypothesis).

    In the coin tossing example, one could sample a new data set that has the same size as the original data, but with $p_H = 0.8$. As a second example, one could take the diffusion model parameters estimated from empirical data and simulate a new RT distribution from those parameters. And as a third example, suppose that each subject provides data (e.g., mean RT on the Stroop task) before an experimental manipulation (coded −1); and also provides data after the experimental manipulation (coded +1). The null hypothesis may state that the experimental manipulation has no effect; therefore, one can under this hypothesis randomly shuffle the (−1,+1) pairs for each subject, and thus simulate a novel data set based on the reshuffled empirical data.

3. Calculate a new statistic based on the simulated data, T^{sim}.

4. Repeat steps 2 and 3 as many times as computationally feasible (e.g., 100,000 times).

5. Based on the many (e.g., 100,000) samples, construct the distribution of T^{sim}.

6. Compare T^{obs} with the distribution of T^{sim} in order to calculate a p-value. If T^{obs} is in the tail of the T^{sim} distribution (i.e., T^{obs} has a low p-value), this suggests a bad model fit.

Such a resampling approach is widely used in cognitive neuroscience because it obviates the need to derive exact distributions (Maris & Oostenveld, 2007).[1]

Model Testing across Modalities

An increasingly popular approach in cognitive neuroscience is to first estimate parameters of a model based on one data modality (e.g., behavioral data), and then test the resulting model in a different modality—typically, a neuroscientific measurement such as electroencephalography (EEG) or functional magnetic resonance imaging (fMRI). The latter then provides an indication of whether the processes postulated by the model can also be identified in neural data. This is a very useful approach because it provides *links* across different levels of investigation. I mentioned in chapter 1 that cognitive neuroscience operates across diverse levels of investigation, and a key goal of our scientific endeavor is investigating how the different levels interact. Model testing across modalities is one approach to tackling this issue. I will illustrate this approach with a few examples.

Model-Based fMRI

In model-based fMRI, model parameters are first estimated, typically based only on behavioral data. In a subsequent step, based on the estimated parameters, trial-by-trial model-based regressors are constructed to be used in the fMRI analysis.[2] Voxels are then sought across the brain that correlate with the model-based regressors. Voxels that survive statistical thresholding, and thus correlate sufficiently strongly with the regressor, are interpreted to implement at least part of the model that generated the regressor.

An application of this method was reported in O'Doherty et al. (2004), which delivered both a classical and instrumental conditioning procedure to subjects in an fMRI scanner. They fitted a temporal-difference reinforcement learning model (a generalization of the Rescorla-Wagner model; see chapter 9) to the behavioral data. Based on the resulting estimated parameters, they then constructed a regressor that tracked value on a trial-to-trial basis. They also constructed a regressor for prediction error and included both in their statistical model for the fMRI data. They found that different parts of the striatum responded to prediction errors in the two tasks (ventral striatum to both tasks; dorsal striatum only to instrumental conditioning). They concluded that one part (ventral) of the striatum calculates the values of states of the environment, and another part (dorsal) calculates the values of actions. This is consistent with the actor/critic distinction proposed in the reinforcement learning literature (see chapter 8).

One potential hazard of model-based fMRI is that model parameter estimation can be very difficult in complex cognitive models (see chapter 6). However, mitigating this point, some authors have argued that estimating parameters is not actually required in model-based fMRI because the regression predictors are highly correlated across different parameter values (Wilson & Niv, 2015). For example, a prediction error–based learning model will generate strong prediction errors on surprising events, and weak prediction errors on expected events, regardless of the exact value of the learning rate (except, of course, for extremely small learning rates).

Another approach to test model predictions at the fMRI level is via a representational dissimilarity matrix (RDM) (Kriegeskorte et al., 2008), a matrix that represents the similarities between specific stimuli or conditions. An especially interesting case arises when similarities are defined according to a specific computational model. In this way, we can query which brain areas consider the same stimuli or conditions similar, as the model finds similar. After construction of the RDM, a whole brain search is performed to investigate which brain areas show the same similarity structure as the model (as instantiated in the RDM). Brain areas that are thus identified are those that find the same conditions similar as the model finds similar.

For example, the model may consider conditions A and B to be very similar to each other, but both conditions A and B very different from condition C. If a brain area also considers A and B similar and C very different from both, this area will be correlated with the model RDM. An area with such a correlational structure is interpreted to implement the same computational operations as the model. An important aspect of the RDM approach is that the brain and the model are never directly compared. One only investigates whether the brain and the model find the same conditions similar (regardless of, say, how basic processing units behave in the two structures). This is an important property because it allows connecting not only across the model and the brain, but also in principle between different data modalities (e.g., EEG and fMRI) or different species (e.g., macaque and human), thus addressing the important linking problem in cognitive neuroscience discussed in chapter 1 (Kriegeskorte et al., 2008).

An example of the RDM approach was reported in Holroyd et al. (2018), a study that asked subjects to perform a multiple-step task (action sequence). They were asked to perform one of four possible action sequences (make either tea or coffee; add water first or second). Subjects made a choice between different actions (e.g., take water, take cream) at every step to implement the sequence. A recurrent network model performed the same tasks (see the discussion of this topic in chapter 5). This was a model of the anterior cingulate cortex (ACC), reported earlier in Shahnazian and Holroyd (2017).

After model simulation, an RDM was constructed, representing the similarities between conditions according to the recurrent ACC model, as measured in the model's hidden unit activations. For example, one entry of the matrix would implement how similar the actions "take water" (in one action sequence) and "take cream" (in another action sequence) are, according to the hidden units in the model. The authors then performed a whole-brain search to investigate which brain areas exhibited the same similarity structure. Such brain areas would presumably implement the model's computations. As predicted, ACC turned out to correlate most strongly to the model-based RDM, suggesting that ACC keeps track of where one is located in a broader action sequence.

Model-Based EEG

A similar approach can be taken with EEG. One first estimates model parameters based on behavioral data in order to correlate them afterward with EEG signals. An advantage of EEG is that activation can be measured at a much faster time scale, so within-trial dynamics predicted by the model also can be compared with the EEG data.

To implement this model-based EEG approach, Collins and Frank (2018) administered an associative learning task and fitted a model that combined a slots-based working memory with a Rescorla-Wagner trial-and-error learner. Then they attempted to decode model quantities (e.g., stimulus value generated by the model), at different time points in a trial. Interestingly, a quantity relevant for the Rescorla-Wagner model (stimulus value) could be decoded earlier in a trial than a quantity that is relevant for the working memory component (set size). This finding supports the idea that Rescorla-Wagner learning processes operate faster than, but partially in parallel with, working memory processes. Next, the authors observed that model-based EEG indices of stimulus value (measured at trial onset) were negatively correlated with model-based EEG indices of reward prediction error (measured at feedback onset). This confirms a prediction of the Rescorla-Wagner model because prediction error = reward – value, leading naturally to a negative correlation between value and prediction error. A similar result was obtained at a between-subjects level by Silvetti et al. (2014): subjects with stronger cue-related value prediction showed less prediction error in rewarded trials during subsequent feedback.

Model Comparison

In many cases, there is more than one model that can potentially fit the data, each implementing a different neural or cognitive hypothesis. Given that no model is actually ever true, a model comparison approach has merits relative to the model testing approach discussed in the previous sections, where the validity of a single model (i.e., is it true or not?) was evaluated. In particular, increasing the number of data will lead to a clearer win for one of the models under consideration rather than a rejection of the single model, as in the model testing approach.

A straightforward way to check which model fits better is to inspect the likelihood of each model at the estimated parameter. The model with the highest likelihood is proclaimed to fit best. However, this approach is problematic because it doesn't consider model complexity. Naturally, more complex models will fit the data better simply because, almost by definition, they have more flexibility (Pitt & Myung, 2002). Consider again the linear regression model. Having more regressors will surely increase a model's fit value [e.g., as measured via the squared correlation (called R^2) between observed and model-predicted data points]. However, a model with more parameters will not necessarily yield better (interpretable, robust) results. A good model balances model fit and complexity.

Testing and Comparing Computational Models 117

Several criteria have been proposed that consider both model fit and model complexity. Perhaps the simplest one is Akaike's Information Criterion (AIC). Here, one simply adds two times the number of parameters to $-2 \times$ log-likelihood:

$$AIC = -2\log L + 2K. \tag{7.2}$$

The likelihood term is calculated using the maximum-likelihood parameters. For model comparison, one picks the model with the lowest AIC, which would correspond to a model that balances high likelihood (L) with low complexity (K). The weighting of L and K in equation (7.2) may look a bit arbitrary; however, AIC can actually be derived from a statistical criterion. Specifically, AIC is an approximation of the "distance" between the observed data and the data predicted by the model, corrected for model complexity (Farrell & Lewandowsky, 2018; Sahani, 1999).

Another criterion for choosing the best-fitting model is the Bayesian Information Criterion (BIC). The logic behind the BIC is that a specific parameter value is not important. Instead, one would like to know the average fit of the model (measured by its likelihood), averaged across all possible parameter values (rather than just in the maximum-likelihood parameters). Each possible parameter value yields a different likelihood, all such possible likelihoods are averaged, and each likelihood is weighted by the parameter value's prior probability.

We'll see in chapter 11, on Bayesian statistics, what prior probability means. For now, just note that it corresponds roughly to how much belief one assigns to a given parameter value under a specific model. To see why calculating the average likelihood (rather than just a single likelihood at the optimal parameter point) takes model complexity into account, consider a model with a high likelihood at the optimal point, but with a large parameter space, where many points yield much worse fits (likelihoods). The average likelihood approach will average both the good and the bad likelihoods, thus "punishing" the model for predicting a bad fit at other parameter values.

In contrast, a model with a good fit at the optimal point but a small parameter space (a less complex model) will less strongly correct for bad fits in other parameter values than the optimal one, and thus receive less punishment for model complexity. Calculating this average likelihood can be an intractable problem when a model has many parameters, but in some instances, it (or more specifically, -2 times the log of the average likelihood) simplifies to the following convenient equation (Murphy, 2012):

$$BIC = -2\log L + K \log N,$$

where N is the number of data points used for estimating the parameters, and the likelihood is calculated as for AIC, using the maximum-likelihood parameters. Again, this criterion consists of a fit term ($-2\log L$) and a complexity term [$K\log N$]. The model with the lowest BIC is considered best. However, note that BIC punishes model complexity much more severely than AIC does: the slope of AIC as a function of model complexity (K) is 2, whereas the slope of BIC as a function of model complexity is $\log(N)$, and usually $\log(N) \gg 2$.

To illustrate the relations between likelihood, AIC, and BIC, consider table 7.1. Minus log-likelihoods are shown in this table, to facilitate interpretation: a lower number always means "better." The data consist of 1,000 coin tosses. In data set 1, a single parameter $p = 0.6$ generates all the coin tosses. In data set 2, a separate parameter generates data in part 1 ($p_1 = 0.5$; trials 1–500) and in part 2 ($p_2 = 0.7$; trials 501–1,000). Model 1 assigns a single (to be estimated) parameter to all trials; model 2 assigns a separate parameter to the first 500 trials and another parameter to the next 500 trials.

First, consider data set 1 (upper part of table 7.1). Obviously, model 2 has the better likelihood. It is more complex and thus fits the data better, if only slightly so. However, model 1 has better (i.e., lower) AIC and BIC values; it is less punished for its complexity than model 2 is. Because the data were, in fact, generated by a single parameter (as model 1 correctly assumes), this result illustrates that AIC and BIC perform as they should.

A quite different picture is seen in data set 2. Here, the data-generation model was actually consistent with the more complex model 2. As one can see, model 2 performs better in terms of likelihood, AIC, and BIC. Thus, the model comparison process can choose the more complex model as being the better one, but only if its extra complexity substantially increases model fit.

AIC and BIC yield a clear model selection criterion: pick the model with the lower criterion value. However, consider again data set 1 in table 7.1. Here, model 1 is the better one (e.g., based on AIC). But how do we know how robustly model 1 is better? Is an AIC of 1,343.0 so much better than an AIC of 1,344.4 that we can safely ignore all consideration of model 2 in future work? Or can a new data set yield a completely opposite pattern? To yield a more interpretable measure, Wagenmakers and Farrell (2004) propose estimating the probability that model i is the correct model, given the data, and given that one of the models under scrutiny is the correct one. They propose the following statistic:

$$weight_i = \frac{\exp\left(-\frac{1}{2}\text{AIC}_i\right)}{\sum_j \exp\left(-\frac{1}{2}\text{AIC}_j\right)}.$$

Table 7.1

$-\log L$, AIC, BIC, and cross-validated $-\log L$ for two data sets for two "coin toss" models

	$-\log L$	AIC	BIC	Cross-validated
Data set 1				
Model 1	670.5	**1,343.0**	**1,344.5**	**674.0**
Model 2	**670.2**	1344.4	1347.3	674.1
Data set 2				
Model 1	674.6	1,351.2	1,352.7	670.3
Model 2	**655.3**	**1,314.6**	**1,317.6**	**644.8**

Note: The best model according to each fit measure is indicated in bold.

One advantage of this transformation into so-called AIC weights is that the resulting fit measures now live on a 0–1 scale, which is more interpretable than the original likelihood scale.

One way of looking at AIC and BIC is that in calculation of the log-likelihood (logL), the data are used twice: to calculate the maximum-likelihood parameters, and also to evaluate the likelihood in those same data. This may cause overfitting, and that is why the complexity term needs to be added. Thus, another way to compare models is to simply calculate the log-likelihood, but on a separate data set that was not used for estimating the parameters. This approach is called *cross-validation*. The logic behind it is that model complexity that does not capture actual structure in the data cannot yield an advantage in fitting another, held-out data set. Indeed, the model could not have tuned its parameter values to the held-out data. Thus, cross-validated log-likelihood takes into account both model fit and model complexity, just like AIC and BIC. Cross-validated log-likelihoods are reported in the last column of table 7.1 for the two coin-tossing data sets. As can be seen, it follows the pattern of AIC and BIC and punishes the model that is overly complex (model 2 for data set 1 in table 7.1).

Next, consider model comparison for the learning model applied to the four-armed bandit scenario discussed in chapter 6. Table 7.2 shows the likelihood, AIC, BIC, and cross-validated likelihoods for two learning models and two data sets. Model 1 assumes a single learning rate for positive and negative prediction errors (i.e., as in all models considered thus far), whereas model 2 assumes a separate learning rate for positive and negative prediction errors (Frank et al., 2005). Each data set consists of 1,000 trials; every 50 trials, all the reward probabilities switch (i.e., as in a probabilistic reversal learning experiment).

In data set 1, identification of the correct model is hard; depending on the index used, either model 1 or model 2 is favored, and the fit indices for the two models are generally similar. In some sense, this is not unexpected: model 2 does actually fit the data, but it is just slightly more complex than necessary. In contrast, in data set 2, a large difference was implemented between positive and negative learning rates (0.8 and 0.05, respectively). In this case, the more complex model 2 is clearly favored by all the fit indices.

Table 7.2
–logL, AIC, BIC, and cross-validated –logL for two data sets for two "4-armed bandit" learning models

	–logL	AIC	BIC	Cross-validated
Data set 1				
Model 1	648.2	1,300.6	**1,310.4**	642.8
Model 2	**647.2**	**1,300.4**	1,315.2	**640.6**
Data set 2				
Model 1	689.7	1,383.4	1,393.2	688.7
Model 2	**683.1**	**1,372.2**	**1,387.0**	**686.2**

Note: The best model according to each fit measure is indicated in bold.

Unlike AIC and BIC, cross-validation doesn't simply count parameters to measure model complexity, but also considers the functional form of the model. Some other fit measures that go beyond counting parameters are discussed in Pitt et al. (2002). In any case, when one is in doubt whether a fit measure (AIC, BIC, cross-validation, or any other) can identify the correct model for a given data set, it is possible to carry out a model recovery study, very similar in spirit to the parameter recovery that I discussed in chapter 6. To evaluate model recovery, one simulates data under two or more models (say, M1 and M2), and then evaluates whether data generated under M1 are also identified as originating from M1 by the fit measure, and likewise for M2. See Wilson and Collins (2018) for further details.

Finally, at a practical level, it is worth considering that we usually collect data in several participants, not just a single one as was implicitly assumed thus far. We also thus want to aggregate fit measures across participants. One way to perform model comparison in this case, is to simply add the fit measure (AIC, BIC, or cross-validated log-likelihood) across subjects and compare the added values for different models. This approach is valid because all these measures approximate a log-likelihood of the data, which can be added across independent measurement units. For details on the procedure, consult Farrell and Lewandowski (2018).

Applications of Model Comparison

Models of Working Memory: The Sequel

In chapter 6, I discussed three popular working memory models (slots, resources, and interference). How can we empirically disentangle them? A very informative study in that respect was reported by van den Berg et al. (2014). The authors systematically formulated models in a factorial design, crossing the nature of precision (four levels; e.g., discrete or continuous), the number of items that can be remembered (also four levels), and whether spatial binding errors (as discussed later in this chapter) occur. Together, this led them to compare AIC values across 32 different models ($4 \times 4 \times 2 = 32$). They found that models that allow continuous variations in precision across set sizes, items, and trials (like the resource and interference models, but unlike the slots model) were clearly empirically favored.

Oberauer and Lin (2017) also compared their own interference model to versions of the slots and resource models. Based on AIC, the interference model systematically fit their data best of all. Let's consider one reason why the interference model outperformed the others. Oberauer and Lin (2017) observed that subjects sometimes reported the locations of other stimuli from the display (spatial binding errors). From an interference model perspective, it is clear why this can occur; context features will (weakly or strongly) overlap for target and nontarget items, so that also nontarget items will be (weakly or strongly) activated upon cue presentation. This is a natural aspect of the interference model, and a general property of network models of memory (see also chapter 3); but it could be added only post hoc in the slots or resource models.

Learning Models of Decision Making

In chapter 6, I mentioned that Hauser et al. (2014) observed different temperatures ($1/\gamma$) in control participants versus juveniles with attention deficit hyperactivity disorder (ADHD), but a similar learning rate across the two groups. However, this is not the complete story. The authors actually fitted two different models to their data—one being the learning and decision making model described before, and a second model where the learning rate could adaptively change between experimental conditions. In particular, in a condition where payoff probabilities p_i occasionally change, it is adaptive to increase one's learning rate (Behrens et al., 2007).

The authors fitted the two models to both control and ADHD groups and compared their fit (using Bayesian model comparison); this is yet another model comparison approach, but it goes beyond the scope of the current discussion. The authors observed that the standard learning model (i.e., with fixed learning rate) fit best in the ADHD group, but the adaptive learning rate model fit better in the control group. Even though the learning rates were overall similar across the two groups, only the control group could adapt its learning rate to the experimental condition they were currently in.

To sum up, the past two chapters have been outliers in this book, in the sense that rather than focus on model building, they focused on the statistical treatment of computational models. In this chapter, I discussed several methods for model evaluation. Currently, a popular approach is AIC or BIC comparison across candidate models because it is broadly applicable (if parameters can be estimated) and takes into account both model fit and model complexity. However, model comparison is ideally complemented by testing model predictions.

8 Reinforcement Learning: The Gradient Ascent Approach

How to climb a mountain? One step at a time.

—Inspirational quote

The importance of reinforcement for learning adaptive behavior was already empha-sized more than a century ago (Thorndike, 1901). In fact, according to the subsequent behaviorist movement, *all* adaptive behavior is shaped by reinforcement (Skinner, 1938); and to procure maximal reinforcement, animals (including humans) can only learn what response to emit in what situation. The behaviorist movement was strongly criti-cized in theory (Chomsky, 1959), and in experiments demonstrating that rodents moving about in a maze can learn the map of the maze, and without obtaining rein-forcement (so-called latent learning) (Tolman, 1948). This ended what some have called a "dark age" in psychology and spurred the cognitive revolution (Miller et al., 1960).

Around that same time, computational scientists developed methods for achieving optimal behavior that were strongly inspired by the behaviorists' rich animal database. Their methods came to be known as *reinforcement learning* (RL) (Sutton & Barto, 2018). RL is currently a very active field of research at the crossroads of cognitive neuroscience and machine learning. Interestingly, it has become increasingly clear in recent years that to solve complex tasks, RL algorithms must combine the insights from both the behaviorist and the cognitive movements: a focus on procuring rewards by learning the right actions, but supported by cognitive maps (Silver et al., 2016).

In the next two chapters, I consider such RL models. However, even when restricting my attention to models in cognitive neuroscience, I cannot do justice to this big and burgeoning field here. I therefore refer interested readers to the reference list at the end of this book should they wish to know more about this topic (e.g., Sutton & Barto, 2018).

The goal of all RL models is to maximize reward. The idea behind this approach is that animals in the wild attempt to obtain as much reward as possible, modulated by cost factors such as risk of loss, punishment, and effort investment. Although easy to define in a computer algorithm (reward = 1 if some goal is met, and zero otherwise), reward is harder to define in real life. Again, the prototypical example of reward would be an animal who finds or receives food. But even food is not universally rewarding; it presents itself as a reward only when the animal in question is not satiated. However, this issue can be solved by formulating the RL problem as a homeostasis problem (Keramati & Gutkin, 2014). *Homeostasis* is a steady state that animals try to reach, which is defined across several physiological variables. For example, animals want to obtain enough food (without overeating), enough water (but not too much), enough warmth (again, not too much), and so on. Obtaining a reward can be reformulated as getting closer to homeostasis. This immediately solves the conceptual problem here: Food is rewarding if the animal is underfed, but no longer if it is satiated. This approach can solve several other conceptual issues; I refer readers to Keramati and Gutkin (2014). Currently, I will simply assume that reward can be unproblematically defined, and agents must obtain as much of it as possible.

An appealing feature of RL approaches is that they use feedback from the environment, but without asking too much from that environment. They only require feedback on whether an action was good. In contrast, supervised methods (as discussed in chapters 3–5) require much more: for each output unit, they require the correct (i.e., the target) value to calculate the relevant prediction errors, referred to as delta (δ) in chapters 4 and 5. Because RL methods demand some supervision from the environment, but not too much, they are sometimes classified as *semisupervised learning methods*. However, because the term *semisupervised learning* is used differently by different researchers (Ravichandiran, 2018), I will not use it.

Recent decades saw an explosion of research on both the computational strength and biological plausibility of RL methods. These aspects will be unpacked in this chapter and chapter 9. This chapter considers a straightforward application of the gradient ascent principle from earlier chapters toward the goal of achieving RL. In particular, our goal function will consist of average reward (across trials), and we will take small steps in the direction of the derivative of this quantity in order to "climb" the reward surface as quickly as possible. Although this approach is rarely used in cognitive models (but see Quax & Van Gerven, 2018), it serves as a bridge between chapters 2–5 (which used gradient-ascent algorithms for optimization) and chapter 9 (which uses another approach toward RL).

At a theoretical level, it is instructive to see that gradient ascent on reward produces quite sensible algorithms, which are in many ways similar to the supervised-learning gradient algorithms from earlier chapters. Moreover, state-of-the-art machine learning algorithms (Mnih et al., 2015) do use aspects of the gradient ascent approach to RL in some cases; this point is discussed in chapter 9.

Gradient Ascent Reinforcement Learning in a Two-Layer Model

Consider a two-layer model like the one shown in figure 8.1a. Input units (x) project in a linear manner to output units. Each output unit i is activated ($y_i = 1$) according to the following equation:

Reinforcement Learning: The Gradient Ascent Approach 125

$$\Pr(y_i = 1) = \frac{1}{1 + \exp\left(-\gamma \sum_j x_j w_{ij}\right)}. \tag{8.1}$$

Based on the action [as instantiated in the binary output pattern $\mathbf{y} = (y_1, \dots, y_I)$], a single, scalar reward signal R is provided to the network (see figure 8.1a). This reward signal can be used to change the weights from the input to the output layer. Note that as in earlier chapters, we have an inverse temperature parameter γ. Here, this parameter controls the amount of exploration versus exploitation; if γ is higher, responding will be more variable and the model will explore its options more (also discussed later in this chapter).

We attempt to maximize the expected value of this reward signal R, averaged across all possible situations $\mathbf{x} = (x_1, \dots, x_J)$ (Williams, 1992). We thus attempt to maximize

$$E(R \,|\, W), \tag{8.2}$$

where an expectation is calculated across all possible inputs $\mathbf{x}$. Just as we applied steepest descent in chapters 3–5 to find the optimal network weights, we can now apply steepest ascent to find the optimal weights in order to maximize equation (8.2). Thus, we would change weights as follows:

$$\Delta w_{ij} = \beta \frac{\partial E(R \,|\, W)}{\partial w_{ij}}. \tag{8.3}$$

We can then calculate the derivative in equation (8.3) and approximate it for online updating (i.e., at each trial n). This leads to the following result (Williams, 1992; also see exercise 8.1):

$$\Delta w_{ij} = \beta \gamma \sum_{\mathbf{y}} \Pr(\mathbf{y} \,|\, \mathbf{x}^n) \times E(R \,|\, \mathbf{y}, \mathbf{x}^n) \times x_j \times (y_i - p_i) \tag{8.4}$$

for trial n, where p_i is shorthand for the probability defined in equation (8.1). An input pattern at trial n is denoted as $\mathbf{x}^n$. Note that in equation (8.4), we take a sum across all

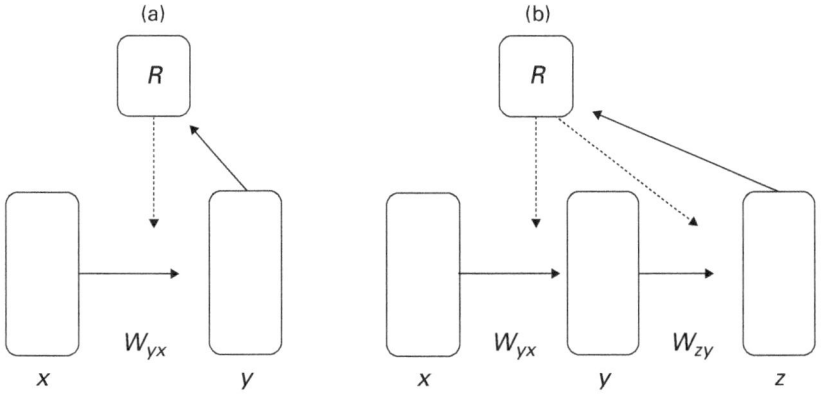

Figure 8.1
Two possible RL architectures: (a) two-layer model; (b) three-layer model.

possible output vectors $\mathbf{y}$ (i.e., across 2^I patterns). Equation (8.4) is very hard to calculate exactly, but fortunately one can approximate it by sampling (from the distributions of $\mathbf{x}$, $\mathbf{y}$, and R) as follows:

$$\Delta w_{ij} = \beta \gamma r x_j (y_i - p_i). \tag{8.5}$$

This quantity is easy to compute on a trial-to-trial basis.

The learning rule [equation (8.5)] bears similarities to rules that have been discussed before. As in the Hebbian rule, a weight change occurs only when there is both presynaptic (x) and postsynaptic (y) activity. Unlike the Hebbian rule, but like the delta rule, there is a prediction error term: an organism must first *predict* (p_i) whether output unit i will be active (i.e., whether it will itself choose option i; but note that it can choose more than one option simultaneously). Then it must compare the actual choice (y_i) with the prediction. Thus, if an action is rewarded ($r = 1$) but the action is highly expected ($y_i - p_i$ is small), little learning will occur.

In going from equation (8.3) to equation (8.4) to equation (8.5), I have paid a price: The update requires sufficient *sampling* from the relevant probability distributions $\Pr(\mathbf{x})$ and $\Pr(\mathbf{y}|\mathbf{x})$ to approximate the correct gradient. This is a general issue in RL, typically formulated as the *exploration-exploitation dilemma* (see chapter 9). In supervised learning rules (chapters 3–5), there is also a sampling issue, but it is much less severe in supervised learning. In particular, also in supervised learning, we (or rather, our environment) often must sample sufficiently across input patterns $\mathbf{x}$ to get an estimate of the goal function. But in RL, there is the additional problem that we must sample across an action (output pattern) $\mathbf{y}$, conditional on an input pattern $\mathbf{x}$. Indeed, in RL, the environment doesn't tell us what the correct action (output pattern $\mathbf{y}$) is! For this reason, the inverse temperature parameter γ appearing in equation (8.1), which determines the relative amount of exploration versus exploitation, is very important in RL. A suitable choice of this parameter allows one to explore the action space appropriately.

Exercises

8.1* Derive equation (8.4) from equation (8.3).

8.2 Argue why equation (8.4) can be approximated by equation (8.5).

An *N*-Armed Bandit

Suppose now that not all output vectors $\mathbf{y}$ are possible, but only the vectors $\mathbf{y}_1 = (1, 0, \ldots, 0)$, $\mathbf{y}_2 = (0, 1, 0, \ldots, 0)$, $\ldots$ are. In other words, we must choose one of the components of $\mathbf{y}$. Each component of $\mathbf{y}$ corresponds to one action, and the actions are mutually exclusive. This is equivalent to the N-armed bandit problem discussed in chapter 6, where action i entails choosing bandit i. Indeed, the pattern $\mathbf{y}_1 = (1, 0, \ldots, 0)$ corresponds to choosing the first bandit, $\mathbf{y}_2 = (0, 1, 0, \ldots, 0)$ corresponds to choosing the second bandit, and so on.

The challenge is to find the best of the N bandits on each trial in order to maximize one's reward. We choose each bandit i ($i = 1, \ldots, N-1$) with the following probability:

Reinforcement Learning: The Gradient Ascent Approach

$$\Pr(y_i = 1) = \frac{\exp\left(\gamma \sum_{j=1}^{J} x_j w_{ij}\right)}{1 + \sum_{k=1}^{I-1} \exp\left(\gamma \sum_{j=1}^{J} x_j w_{kj}\right)}.$$

Because environmental features (x_j) determine which bandit will be chosen, this is sometimes called a *contextual bandit problem*. As an example, suppose that you are in a casino, and certain contextual features of a bandit (e.g., shape, size, and location) determine its probability of payoff. You can then learn, based on these features, which bandit to choose. In this case, the update rule, starting from equation (8.3), is very similar to what we have seen before:

$$\Delta w_{ij} = \beta \gamma \sum_{k} \Pr(y_k = 1 \mid x^n) \times E(R \mid y_k = 1, x^n) \times x_j \times (y_i - p_i). \tag{8.6}$$

On a trial-by-trial basis, again sampling from the appropriate distributions, this weight update equation becomes

$$\Delta w_{ij} = \beta \gamma r x_j \, (y_i - p_i) \tag{8.7}$$

for the chosen bandit i. When no features (x) are available and the organism can only see the bandits themselves for each trial, each weight w_i corresponds to the "value" of the bandit i. In this case, the weight update equation simplifies to

$$\Delta w_i = \beta \gamma r (y_i - p_i).$$

Performance of this algorithm on a four-armed bandit problem is shown in figure 8.2 (convolved with a small time window and averaged over twenty simulations) for different levels of exploration γ. Performance is best for intermediate γ values (.8) that balance exploration and exploitation (note that 80% is the highest possible payoff). With a high level of exploration ($\gamma = .01$), performance remains noisy (exploratory) even after finding the best bandit; with a low exploration level ($\gamma = 5, 10$), the system often becomes stuck in a suboptimal solution.

Exercises

8.3　In the N-armed bandit task, what is the probability of choosing the Nth (i.e., the last) bandit?

8.4　In figure 8.2, there is an optimal (intermediate) γ. Do you think that there is also an optimal intermediate learning rate parameter β?

8.5　With low γ, the model is too exploitative; with high γ, it is too explorative. Can you find a way to improve the algorithm and be exploitative or explorative at the right moment? Implement it and compare the performance to figure 8.2.

A General Algorithm

A more general gradient-ascent algorithm for reward maximization called REINFORCE was derived by Williams (1992). This algorithm updates weights based on

$$\Delta w_{ij} = \beta_{ij}(r - b_{ij})e_{ij}, \tag{8.8}$$

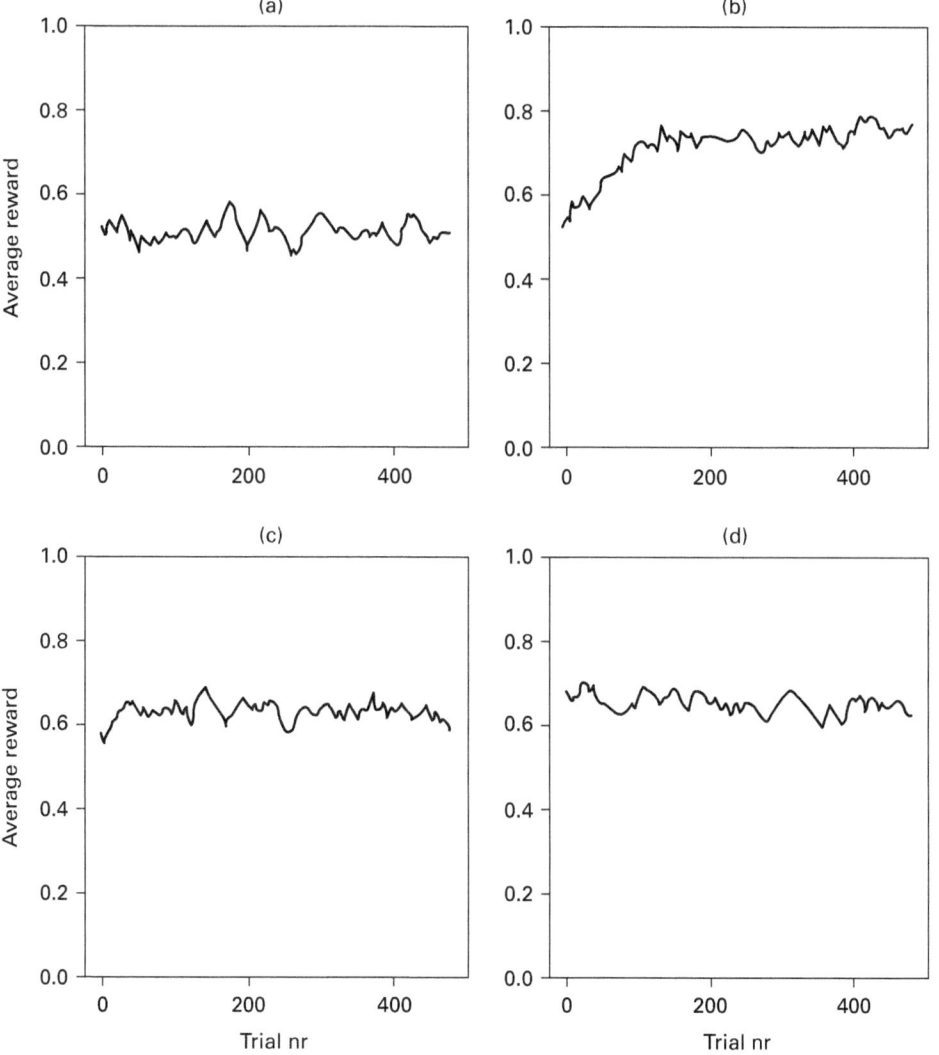

Figure 8.2
The effect of gamma on performance (average reward reaped) in the four-armed bandit problem. Gamma = 0.01, 0.8, 5, and 10 in panels (a)–(d), respectively.

where β_{ij} is a stimulus- and action-specific learning rate. The parameter b_{ij} is a stimulus- and action-specific baseline against which reward r is compared. As a consequence, if reward is higher than this baseline, the weight update will be positive. Further, e_{ij} is a stimulus- and action-specific eligibility trace (roughly, how eligible or ready weight w_{ij} is for being updated), defined as $\partial \log(\Pr(y_i = 1))/\partial w_{ij}$, for any probability function (i.e., not just the one defined in equation (8.1)). However, with the logistic probability function from equation (8.1), the eligibility trace is $e_{ij} = x_j(y_i - p_i)$ (see Williams, 1992), so if we assume this and further set $\beta_{ij} = \beta$ and $b_{ij} = 0$, then the REINFORCE rule is exactly the same as the learning rule [equation (8.7)]. Another useful application for an eligibility trace is to help bridge between a stimulus and reward that appears only later, at a time when the stimulus is no longer physically present (for details, see Williams, 1992).

Williams (1992) demonstrated that the weight update vector in equation (8.8) is positively correlated with the gradient vector. Thus, the REINFORCE algorithm generally ascends the reward function surface. So the algorithm doesn't exactly follows the gradient vector; but for this price, one obtains a larger generality in equation (8.8). In particular, just as in Hebbian learning (chapter 3) and supervised learning (chapters 4 and 5), contrasting an update factor to some baseline b_{ij} can be computationally advantageous. For example, one practical problem with RL is that the weight updates may show strong variability across trials due to the variance of reward (r) across trials. By contrasting the reward to an appropriate baseline (b), this variability can be reduced.

Backpropagating RL Errors

It is also possible to derive a multilayer (or deep) steepest ascent algorithm for the goal function $E(R|W)$ (Williams, 1992). Consider the model structure in figure 8.1b. The input units project to the hidden layer in a deterministic manner:

$$y_i = \frac{1}{1 + \exp\left(-\sum_j x_j w_{ij}\right)}.$$

The hidden units project to the output layer, where an action pattern (now denoted $\mathbf{z}$) is chosen according to a probability function $\Pr(\mathbf{z}|\mathbf{y})$ as before:

$$\Pr(z_i = 1) = \frac{1}{1 + \exp\left(-\gamma \sum_j y_j w_{ij}\right)}.$$

Then the (sampling version) update rule for the hidden-to-output (*j*-to-*i*) weights becomes

$$\Delta w_{ij} = \beta \gamma r x_j (y_i - p_i),$$

and the (sampling version) update rule for the input-to-hidden (*k*-to-*j*) weights becomes

$$\Delta w_{jk} = \beta \gamma r x_k y_j (1 - y_j) \sum_i w_{ij} (z_i - p_i). \tag{8.9}$$

Note that this weight update rule is very similar to the backpropagation update rule from chapter 5, where the prediction error at the output level ($z_i - p_i$) is propagated toward the units at the hidden level, weighted by the same weights w_{ij} that are used in the forward activation pass.

Three- and Four-Term RL Algorithms: Attention for Learning

Although, as previously noted, the current gradient-ascent RL framework is not frequently used in cognitive neuroscience, some models have used similar learning rules. Consider the three-factor rule of RL (Ashby et al., 2007). According to this rule, an action should be reinforced by connecting the representation of the action to the representation of the stimulus (or situation, or state) (factor 1) that led to that action (factor 2), but only when a reward is present (factor 3). The learning rule [equation (8.7)] is exactly of this type.

Biological data in the striatum support such a learning rule (Reynolds et al., 2001). Here, the reward is signaled by dopamine, originating in mesencephalic (i.e., from the midbrain) projections to the striatum. In other implementations of the three-factor rule (Ashby et al., 2007; Schultz et al., 1997) the reward factor rather than the action factor, as in equation (8.7), appears in a prediction error format; I will consider reward prediction errors more extensively in chapter 9.

Deep RL algorithms analogous to equation (8.9) have also been proposed by researchers such as Roelfsema and van Ooyen (2005) and Rombouts et al. (2015). Specifically, these authors developed a four-factor RL rule: synaptic weights change as a function of presynaptic activity, postsynaptic activity, reward prediction error, and finally attention (factor 4). This attentional signal makes sure that the right synapses are changed; and the attentional signal itself is also learned via RL principles. These authors motivated their learning rule from a different angle—namely, by showing that on average, their learning rule leads to similar updates as backpropagation (chapter 5). Yet, note that equation (8.9) is also a four-factor rule: It is a multiplicative function of presynaptic activity, postsynaptic activity, reward, and a term $\sum w_{ij}(z_i - p_i)$ which can be interpreted as an attentional term. If the unit i is active and (unexpectedly) leads to reward, and the hidden unit j contributed to that event (via w_{ij}), this summation term will make the hidden unit j eligible for weight change (from input unit k to hidden unit j). One difference with the Roelfsema and van Ooyen (2005) algorithm is that the attentional term is symmetric in the currently derived gradient ascent algorithm; the same weight is used for message passing from j to i as for increasing eligibility from i to j.

It is quite interesting that the psychological construct of attention can be given a computational rationale in an RL framework. Moreover, there is extensive biological and psychological evidence for reward-attention interactions in the way predicted by such learning rules (Roelfsema et al., 2010). Note finally that the attention factor is similar to the credit assignment factor in the original backpropagation rule (chapter 5), which identified which output units had to change its activation, and which hidden unit weights had to be adapted for that purpose.

As already mentioned, the approach in the current chapter is not very popular in the cognitive neuroscience literature; it is, however, very important in modern artificial intelligence (AI), where it is referred to as *policy gradient* (Peters & Schaal, 2008). Indeed, in the current approach, one directly changes an agent's policy by walking up a reward gradient (I will unpack the concept of a policy in chapter 9). One important policy gradient algorithm is the actor/critic model; here, one part of the model (the actor) computes the relevant policy (as described in this chapter), while another part (the critic) separately learns and evaluates whether an action was good (i.e., of high value) or not. The information from the critic is then used for training the actor (Sutton & Barto, 2018). The actor/critic can be seen as a generalization of the REINFORCE algorithm, where the critic learns a more powerful reward term than $r - b_{ij}$. Several extensions of this basic approach have been proposed in recent years; for an overview (and useful Python computer code), see Ravichandiran (2018).

To sum up, in this chapter, I have demonstrated how one can apply the principle of gradient ascent directly to a model's actions (output) and thus obtain a policy for gaining as much reward as possible. In a two-layer architecture, this led to a simple and intuitive learning rule that combined a Hebbian (input and output activation) with a reward component. In a multilayer architecture, this led to a model with the same factors, but also a factor that could be interpreted as implementing attention to specific hidden units. As already mentioned, this chapter serves as a bridge toward chapter 9, which also covers RL, but the optimization principle is implemented differently than via straightforward gradient ascent on the reward. So let's now cross that bridge.

9 Reinforcement Learning: The Markov Decision Process Approach

Those who cannot remember the past are condemned to repeat it.
—George Santayana

In this chapter, I turn to another approach toward reinforcement learning (RL): the Markov decision process (MDP). This approach has its roots in the optimal control literature from engineering, and it has been very popular in cognitive neuroscience. This is partly because aspects of the MDP approach were found to correspond closely to neurophysiological data in both nonhuman (Schultz et al., 1997) and human (Seymour et al., 2004) primates, and partly because the framework turns out to be quite computationally powerful (Mnih et al., 2015). In this sense, the framework has both computational and neurophysiological credibility.

A classical treatise on this RL approach was done by Sutton and Barto (2018). Readers who want to know more about this approach are advised to consult that book. At a broad conceptual level, the current approach has strong similarities to the RL approach sketched out in chapter 8. First, the *goal* of maximizing reward is unchanged; only the method for achieving that goal is different. Second, the type of feedback received from the environment is similar. The RL learner receives more feedback than in unsupervised learning (which will be discussed in chapter 10), but less feedback than in supervised learning (chapters 3–5). Arguably, the type of feedback in RL models ("You're doing well" or "You're doing not so well") is more similar to the feedback that an actual biological agent usually receives during learning than the detailed feedback target patterns that are assumed in supervised learning. See also the discussion in chapter 8 on the relation between reward and homeostasis. Third, RL is very much action-oriented. RL can learn only by interacting with the environment; it learns by doing. Again, it might be argued that this situation corresponds faithfully to the learning task faced by a biological agent, and a large body of empirical literature shows the importance of action to learning. For example, in human subjects, simply picking a stimulus (i.e., a minimal action) helps to better remember that stimulus (Murty et al., 2015). And the fact that

134   Chapter 9

it learns by acting also makes it a useful approach for training autonomous robots. Of course, the learning algorithms discussed in chapters 4 and 5 are not entirely passive; the organism must make a prediction in order to learn. But still, unlike with RL, they must passively wait for an input pattern to be presented to them.

The MDP Formalism

To understand the MDP concept, look at figure 9.1, in which we consider an agent interacting with its environment. The concepts of "agent" and "environment" are deliberately very general so that one has significant freedom to apply it to one's context of interest. Examples would be a cat (agent) in my home (environment) or a snake (agent) in a jungle (environment). But the concepts can be stretched considerably to encompass a visual cortex (agent) in a brain (environment); a daughter (agent) in a family (environment); a football team (agent) in its football league (environment); and other scenarios.

In each of these cases, the agent receives states and a reward from the environment. As usual, time is divided into discrete time steps indexed t, and so states and rewards are provided as s_t and r_t. In turn, the agent provides actions a_t to the world in the hopes of changing that world (and convincing it to maximize the agent's reward provision). Note here yet another important difference with previously discussed approaches, including the one from chapter 8: states and actions (and rewards) are not coded as sets of distributed features; rather, they are coded in a discrete manner—either a state (or action) is present or it is not. This property makes formal derivations easier, but it renders generalizations (for the agent) problematic. In principle, if two states of the world are even minimally different (consider two cloudy skies that differ by just one cloud), they could be coded by different states, and one would not be able to generalize what one has learned from one state to the other.

So why does one call this an MDP approach? First, a statistical process is a collection of variables $\{X_t\}$ indexed by time t. For example, if the weather can be either rainy or sunny, then the sequence of weather states at subsequent days {Rainy, Sunny, Sunny, Sunny, Rainy, . . .} is a statistical process, or just *process* for short. Similarly,

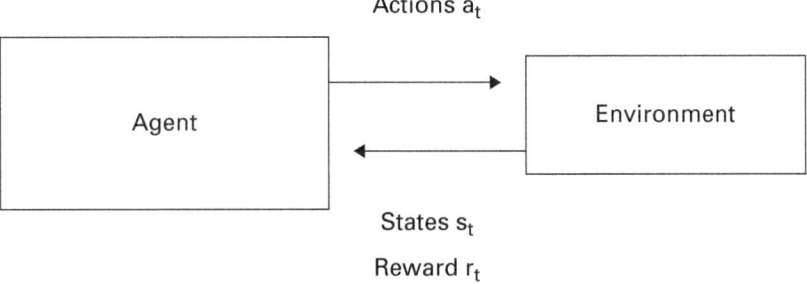

Figure 9.1
Conceptual overview of an agent–environment interaction.

Reinforcement Learning: The Markov Decision Process Approach   135

when repeatedly tossing a coin, the sequence of states {Heads, Tails, Tails, . . .} is also a process. As an aside, this is why this textbook is about modeling cognitive *processes*: we are interested in capturing the dynamics of cognition at both fast and slow time scales. Statisticians say that a process $\{X_t\}$ has the Markov property if its probability distribution at time step t depends only on the state of that process at time step $t-1$. More formally, a process $\{X_t\}$ has the Markov property if

$$\Pr(X_t = x_t | X_1 = x_1, \ldots, X_{t-1} = x_{t-1}) = \Pr(X_t = x_t | X_{t-1} = x_{t-1}).$$

This defines a Markov process; if states are intermixed with actions (from the agent), the process is called a Markov *decision* process (hence MDP).

Exercises

9.1 Does the process in the weather example have the Markov property?

9.2 Does the process in the coin-tossing example have the Markov property?

9.3 Does finding your way through a maze have the Markov property?

After working your way through the exercises, consider another example. Suppose that you're a goalkeeper looking at a player about to take a penalty kick on your goal. If the state of the player (what she is thinking about, where she is standing, with the sun shining in her eyes or not, and so on) is all you need to know in order to estimate where she will shoot, then this process has the Markov property. However, if it matters what happened in earlier states (e.g., whether the penalty occurs in the first or second half of the match), then the process is not Markov. A Markov process has the very useful property that the whole history of the process is summarized in the previous state. As a consequence, given that one knows the previous state, states before the previous one, are irrelevant. This property is exploited in the MDP approach toward RL.

As mentioned, actions are considered to be discrete (an action is executed or not), but otherwise the concept of action is very flexible. Possible actions for the goalkeeper are "jump left," "jump right," "jump up," "stand still," and so on. As in the previous chapter, the central aim of RL is to choose actions that maximize reward. For this purpose, the agent constructs a policy (i.e., a mapping from states to actions) that specifies for each state the corresponding action. This mapping can be probabilistic: for example, a policy could say, "When the shooter runs to the ball in a straight line, I stand still with probability .6 and jump left and right with probability .2 each."

Because of the Markov property, we can specify an agent's policy as a function of the current state s and action a as

$$\Pr(A = a | S = s) = \pi(s, a).$$

This is the probability of choosing action a when one is in state s. If all probabilities $\pi(s, a)$ are either 0 or 1, then the policy is deterministic.

We can represent a policy as a table where rows represent states, columns represent actions, and cells of the table contain the probability of emitting action a when one is in state s. Again, we can do this thanks to the Markov property. Without this property, the whole history $S_{t-1}, A_{t-1}, S_{t-2}, A_{t-2} \ldots, S_1$ would have to be tabulated. Given that the number of states and actions increases exponentially with increasing distance from the present time, doing that would be quite impractical. Because a policy can be represented by a table, we call the MDP approach a *tabular* approach to RL.

Even with a Markov approach, typical tasks are intractable when each state and action is modeled as a discrete event. For example, chess has approximately 10^{45} states; clearly, a table with that number of rows would be hard to construct. Later in this chapter, I consider how researchers have dealt with this problem.

As an example, consider the gridworld from figure 9.2a (partially derived from Sutton & Barto, 2018). There are $5 \times 5 = 25$ states. At each position, the agent has four actions at her disposal (go left, right, up, or down), with the restriction that she does not leave the gridworld. Whenever the agent ends up in state A, she is warped to state A′ and receives a reward of 10 at the same time. Similarly, when the agent ends up in state B, she is warped to state B′, and at that time she also receives a reward of 5. A movement has no cost, but an attempt to move out of the grid has a cost of −1. The agent's task is to find an optimal policy $\pi(s, a)$; that is, in each state s, she must find out what the best action a is.

(a)

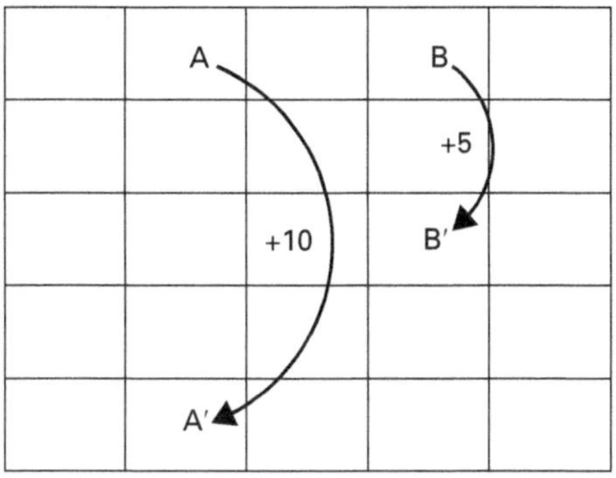

(b)

0.7	0.2	0.5	0.5	0.1
0.1	0.7	0.2	0.3	0.9
0.4	0.4	0.0	0.7	0.1
0.7	0.1	0.7	0.8	0.9
0.0	0.3	0.4	0.5	0.1

(c)

3.2	9.5	4.5	5.3	1.3
1.2	2.9	2.2	1.8	0.4
−0.1	0.8	0.6	0.5	−0.3
−0.7	−0.1	0.0	−0.1	−0.7
−1.3	−0.7	−0.6	−0.7	−1.2

(d)

3.4	8.9	4.5	5.4	1.6
1.6	3.1	2.3	2.0	0.6
0.1	0.8	0.8	0.4	−0.3
−0.9	−0.3	−0.3	−0.5	−1.1
−1.8	−1.2	−1.1	−1.3	−1.9

Figure 9.2
A gridworld and its values if the agent follows a random policy. States are numbered from left to right and from top to bottom. (a) Gridworld. (b) Initial values of the states. (c) Estimates after five iterations. (d) Final value estimates. Panel (a) was reproduced with permission from Sutton and Barto (2018).

Reinforcement Learning: The Markov Decision Process Approach 137

To find such an optimal policy, the MDP approach uses the important concept of *value*. The value of a state *s* in policy π is defined to be the average reward that one would obtain if one starts in state *s* at time *t* (i.e., s_t), and then follows policy π thereafter. The first reward received after s_t occurs at time step $t+1$ and is denoted as r_{t+1}. Rewards obtained in the future are typically less enticing than rewards received right away. For example, if given the choice between receiving 100€ right now or 100€ in a month, most people would prefer the first option. This idea is formalized by assuming that rewards after time step $t+1$ are discounted by a discount parameter $\eta < 1$. Thus, the value of state *s*, given that one follows policy π, is written as follows:

$$V_\pi(s) = E\left(\sum_{k=0}^{+\infty} \eta^k r_{t+k+1} \middle| s, \pi \right). \tag{9.1}$$

To see the meaning of the discount parameter η, consider the case where $\eta = 1$; here, all rewards are of equal importance, no matter how far in the future they are. A person with $\eta = 1$ is indifferent between receiving 100€ right now or receiving it next month. In contrast, if $\eta = 0$, then only the reward r_{t+1} at the next time step is relevant (defining $0^0 = 1$). Of course, at the start of learning, the value $V(s)$ is not known. Figure 9.2b shows an initial random value distribution over the 25 states of the gridworld. I will discuss how these values can be estimated in the next section.

In addition to the value of a state, one can calculate the value of an action. In the gridworld example, the value of an action could refer to the value of going leftward, given that one is in a particular starting state. More generally, we will consider the value (i.e., average reward) of choosing action *a*, given that one is in state *s* and one follows policy π afterward. This is typically denoted as $Q_\pi(s, a)$, and is defined as

$$Q_\pi(s, a) = E\left(\sum_{k=0}^{+\infty} \eta^k r_{t+k+1} \middle| s, a, \pi \right).$$

Note that *V*(.) for any policy is most naturally represented as a vector, with the number of elements equal to the number of states; and *Q*(.) is represented as a matrix (or table), where rows represent states (*s*) and columns represent actions (*a*). In figure 9.2b, the *V*-values are represented as a matrix because the world is two-dimensional (2D), but we can easily imagine (numpy.reshape) this as a vector with 25 elements.

The point of estimating *V* or *Q* (or both) is to use those estimates to implement a better policy π. Each approach (estimating *V* or *Q*) has its own advantages and disadvantages. The good thing about estimating the *Q*-matrix is that it requires no extra knowledge at the time of action. For each state *s*, one can simply pick the action *a* that maximizes the expected reward (i.e., *Q*). The drawback of *Q* is that it's a larger structure than *V*: *Q* is of size (number of states × number of actions), whereas *V* is only of size (number of states). Thus, storing *Q* requires more memory. The disadvantage of calculating and using *V* for navigating the world, on the other hand, is that it requires having extra knowledge at the time of action. Indeed, because we have to reach for the values *V*(*s*) from each potential action *a*, this approach (of using *V* for choosing an action) requires predicting the subsequent states *s*, given that one performs action *a*.

As an example, consider the gridworld again. Here, the environment is so simple that there is no clear difference between estimating V and Q. However, consider the slightly more complex slippery gridworld, in which there is a slipping probability p that you move to a different state than intended (which is randomly chosen from the available states). Thus, the probability of going toward state 6, given that one starts in state 7 and chooses action $a =$ Left, is only $1 - p$. In this case, the V matrix is not sufficient to choose an appropriate action. One must also consider the slipping probability p to make a good choice (action). Instead, if the Q matrix was available, this p would not be required. One could simply pick the reward-maximizing action a again.

Finding an Optimal Policy

Given either V or Q, one can implement a policy. But we can also use those quantities to find a better policy. In fact, the final goal of MDP RL is to construct an optimal (i.e., reward-maximizing) policy, just as in the gradient ascent approach from chapter 8. Specifically, we seek a policy π^* such that

$$V_{\pi^*}(s) \geq V_\pi(s)$$

for every state s and for every policy π. This is quite a strong requirement: such a policy π^* would guarantee that in *every* state s, the value you would get by following π^* is at least as large as you would get by following any other policy π. In particular, if such a policy π^* should be found, it would be a much stronger result than what was obtained in the RL gradient ascent approach of chapter 8. There, we "merely" aimed to act in order to reach a (possibly local) maximum of the average reward function across all states. But in the MDP RL approach, we attempt to find a policy that is the best one in each possible state s. The MDP assumption guarantees that such a policy exists and that it can be found.

To find this optimal policy π^*, we alternate taking two steps. We start from an arbitrary policy π. For this policy, we estimate the corresponding value functions $V_\pi(s)$ (for stimuli) or $Q_\pi(s, a)$ (for actions). This step is called *value estimation*. Once the values (V or Q) for a given policy are found, we aim to find a better policy based on those estimates. This is called *policy updating*. However, once the policy is changed, the values V (or Q) also must be changed. Indeed, the values depend on the implemented policy π, as is clear from equation (9.1). Thus, the two steps (value estimation and policy updating) can be repeatedly alternated. I now describe these two steps in more detail.

Value Estimation

Three methods are popular for value estimation: dynamic programming, Rescorla-Wagner, and temporal differences. These are all described next.

Dynamic Programming

The first approach to value estimation requires that we know what happens in the world after we do something; in other words, that we have a *Model* of the world. It is

Reinforcement Learning: The Markov Decision Process Approach 139

very important not to confuse the word *Model* as used here with the use of that term used elsewhere in this book. As used in this discussion, *Model* refers to a structure in the brain of the acting agent; elsewhere in this book, *model* refers to a structure in the brain of the researcher. To avoid confusion, I could have used the word *map* instead of *Model*, but it is advantageous to use the term *Model* here, as several important studies on Model-based RL refer to this exact use of the word *Model*, meaning a structure in the agent's brain (Daw et al., 2005; Otto et al., 2013). When reading this literature, keep in mind the difference between the two meanings of the word *model*. To emphasize the difference in this discussion, I capitalize this word when I refer to a structure in the head of the agent.

To be more specific, the Model that the dynamic programming approach requires consists of the transition probabilities $\Pr(S_t = s' \mid S_{t-1} = s, A_{t-1} = a)$, or $\Pr(s' \mid s, a)$ for short. This is the probability of going toward state s', given that one has just performed action a in state s. The slippery gridworld with slipping probability p is one example of such a Model. Another example of part of such a transition matrix is shown in figure 9.3b, applying to the rodent maneuvering its way through the maze shown in figure 9.3a.

If an agent has this transition matrix at her disposal, she can apply dynamic programming to find the values of states and actions as follows. Specifically, dynamic programming entails that, due to the Markov assumption, it is possible to write the values $Q_\pi(s, a)$ as a linear system of equalities (Sutton & Barto, 2018):

$$Q_\pi(s, a) = \sum_{s'} \Pr(s' \mid s, a) \left(r(s') + \sum_{a'} \pi(s', a') Q_\pi(s', a') \right).$$

This set of equations are called the *Bellman equations* (after the American physicist Richard Bellman). They can be iteratively solved. In particular, one plugs in a set of values

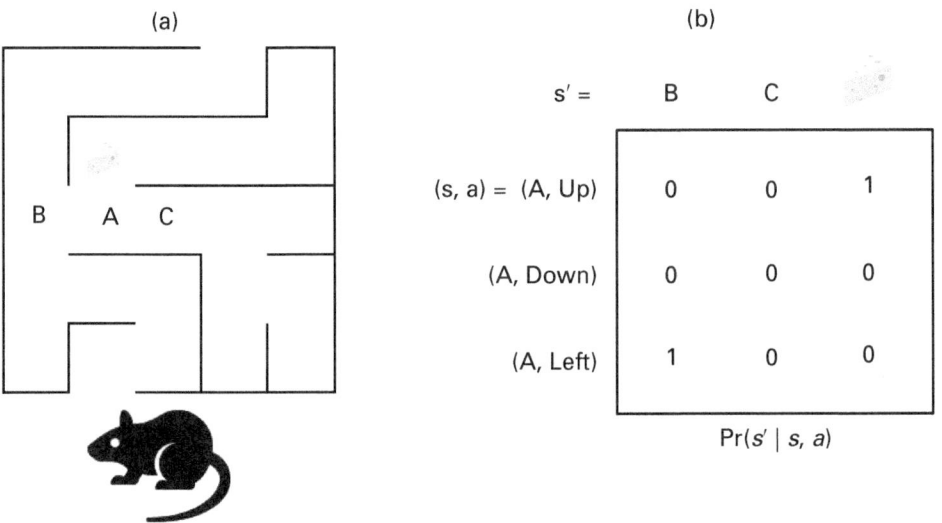

Figure 9.3
(a) Maze. (b) Some of the transition probabilities in this maze.

Q^n, and out comes a new set Q^{n+1}. Thus, as the quote at the start of the chapter suggests, "novel" solutions (Q^{n+1}) are based on older memories (Q^n). This is actually a general feature of dynamic programming methods: novel solutions are iteratively found by combining older solutions.

After convergence, one has the Q-values of policy π. Alternatively, one can write a similar system of equations for the V-values of that policy. This set of equations can also be iteratively solved. After convergence, one has the V-values corresponding to policy π.

Figure 9.2c shows the resulting V-values after five iteration steps for the random policy (i.e., where each allowed action is equally likely). Figure 9.2d shows the final V value estimates after convergence for that policy. One can see that the values in figure 9.2c are intermediate between the initially random values from figure 9.2b and the final estimates in figure 9.2d.

Exercises

9.4 Formulate for yourself as clearly as possible what a specific entry in the grid in figure 9.2d means.

9.5. The discount parameter for generating the data in figure 9.2 is set at $\eta = 0.9$. What would happen to the final value estimates in figure 9.2d if you set it at $\eta = 0.1$? Check your answer using the online code.

Rescorla-Wagner Approach

Dynamic programming has some obvious disadvantages. First, as already stated, it requires knowing the transition probabilities $\Pr(s'|s,a)$ in the world (i.e., it requires a Model of the world). Second, dynamic programming requires the iterative solution of a large set of equations (the Bellman equations).

A more feasible approach is based on Monte·Carlo (i.e., random) sampling. Here, the agent interacts with the environment to obtain samples (s,a,r) from the environment. Intuitively, one queries based on the environment, "If I perform action a in situation s, what reward r do I receive?" With this information, Q-values can be updated on each trial t as follows:

$$Q_\pi(s_t, a_t) \leftarrow Q_\pi(s_t, a_t) + \beta\left(r_{t+1} - Q_\pi(s_t, a_t)\right). \tag{9.2}$$

Both here and later in this chapter, the $\leftarrow$ symbol is the assignment operator, similar to the equals sign as used in computer languages (e.g., in Python, the statement $a = 3$ means "Assign the value 3 to variable a"). A similar approach as in equation (9.2) can be followed to estimate the V-values of a policy π.

Note that the general form [equation (9.2)] is very similar to the delta rule from the supervised learning approach. Indeed, reward r can be considered as a target value to train the value system. Based on the resulting prediction error (delta), the Q-value is updated. Fittingly, this approach is named after two psychologists who emphasized the role of prediction error in learning: Robert Rescorla and Allan Wagner (see the discussion in chapter 4).

Reinforcement Learning: The Markov Decision Process Approach

So isn't MDP RL a type of supervised learning, after all? Note that the estimation part of MDP RL is just the first step in the MDP RL approach. The final goal of RL is not estimation, but rather finding a good (i.e., optimal) policy. Value estimation is just an auxiliary first step. In contrast, in supervised learning, estimation (of the correct outputs) is the final goal. Another important difference with supervised learning, as mentioned before, is that what is required from the environment is much weaker in RL. In supervised learning, every output unit receives the exact information (target) about what it was supposed to do. In contrast, in the RL setup, the environment merely has to say whether a response was good ($r = 1$) or not ($r = 0$).

Compared with dynamic programming, Rescorla-Wagner learning requires much less resources. Most notably, it does not require that the agent knows the transition probabilities $\Pr(s'|s, a)$ in the environment (a Model of the world); it merely requires that the agent can interact with the environment and obtain samples (s, a, r). For this reason, we call Rescorla-Wagner learning a Model-*free* approach. Again, the Model is here used from the perspective of the agent; the agent does not have a Model [i.e., does not possess the quantities $\Pr(s'|s, a)$], and in this sense, this approach is Model-free. On the downside, because it uses less information per trial, Rescorla-Wagner learning is typically less efficient and thus requires more trials to obtain stable estimates relative to dynamic programming. Indeed, consider trying to learn to play chess using Rescorla-Wagner learning. If one considers the act of winning the game as a reward (and intermediate moves as neutral), then only at the very end of the game would you receive information about the usefulness of your moves. But you have made tens (or even hundreds) of moves during the game! How can you appropriately assign the reward at the end of the game to each of the moves (actions) you made? This is again the credit assignment problem that we discussed in earlier chapters.

Exercise

9.6 Strictly speaking, the update rule [equation (9.2)] is valid only when $\eta = 0$. Can you see why?

The Temporal Differences Approach

Dynamic programming is efficient because it does not require actual rewards from every trial in order to learn. Rescorla-Wagner learning, on the other hand, is feasible because it does not require complicated updating algorithms. Temporal difference learning combines the advantages of these two approaches. In the Rescorla-Wagner case, the target for the Q-values is the next-time step reward, r_{t+1}. In temporal difference learning, the target is the current reward *plus the (discounted) value of the next state*. In particular, the temporal difference learning rule for V values is written as

$$V_\pi(s_t) \leftarrow V_\pi(s_t) + \beta\big(r_{t+1} + \eta V_\pi(s_{t+1}) - V_\pi(s_t)\big). \tag{9.3}$$

Note what happens here: The value of the state that was active at time step t (i.e., s_t), is updated using the reward at the next time step (r_{t+1}), but also, using the estimate of the value of the subsequent state, the organism finds itself in $(V(s_{t+1}))$. Thus, even though the value at the next time step $(t+1)$ may not yet be well known, its estimate is nevertheless used as a target for updating the value of the previous state (s_t), whose value is likely even less well known. This procedure is called *bootstrapping*,[1] named after the act of jumping across a fence by pulling the straps around one's boots. This is obviously impossible, and it likewise could look as though temporal difference learning is impossible too.

One might rightfully wonder: how could one learn from imperfect estimates $V(s_{t+1})$? In practice, though, introducing the temporally delayed $V(s_{t+1})$ into the target part of the update rule turns out to be a very efficient way to span the often-long delay between states and rewards. In earlier chapters (e.g., chapter 5), I mentioned the credit assignment problem; there, the problem was to discover which units were to be credited for some error. One can consider this as a *spatial* version of the credit assignment problem. Instead, the temporal difference algorithm is one way to tackle a *temporal* variant of the credit assignment problem. Here, the question concerns how to appropriately award credit to units that were active long before the actual reward. This can be achieved by including the value at the next time step in the update [e.g., equation (9.3)].

As an example, consider the cheese state in figure 9.3. In the Rescorla-Wagner approach, the reward attached to this state can be transferred to state A because state A appears right before the cheese state. However, because the Rescorla-Wagner approach learns exclusively via rewards, the acquired value of state A could never trickle down to state B or C. In contrast, using temporal differences, the updates for states B and C are partially based on the value of state A, and therefore, as soon as state A has acquired some value, this value can be transferred to state B or C. From B and C, the values can subsequently trickle down to their own neighboring states, all the way up to the state where the mouse initiates its journey.

When applied to Q-values, the temporal difference learning rule becomes

$$Q_\pi(s_t, a_t) \leftarrow Q_\pi(s_t, a_t) + \beta\left(r_{t+1} + \eta Q_\pi(s_{t+1}, a_{t+1}) - Q_\pi(s_t, a_t)\right). \tag{9.4}$$

This is called the SARSA rule because of the sequence of elements on the right side of equation (9.4) (s, a, r, s, a). Many other value estimation methods have been described; we again refer interested readers to Sutton and Barto (2018).

One final approach for updating values in RL, is called *Q-learning* (Watkins & Dayan, 1992). Here, one updates a Q-value based not on the action taken (i.e., as in standard temporal-difference learning), but on the *best possible* action that could have been taken at time $t+1$:

$$Q_\pi(s_t, a_t) \leftarrow Q_\pi(s_t, a_t) + \beta\left(r_{t+1} + \eta \max_a\{Q_\pi(s_{t+1}, a)\} - Q_\pi(s_t, a_t)\right). \tag{9.5}$$

Q-learning is called an *off-policy RL algorithm* because one estimates the values for a policy (here, the optimal policy) other than the policy that one is currently following. Instead, Rescorla-Wagner and temporal differences are called *on-policy RL algorithms:* one attempts to estimate the values of the policy that one is currently following.

Policy Updating

Once the value of a policy (either Q or V) is estimated, the policy can be updated. In particular, a new policy is constructed based on the newly minted Q matrix or V vector. For Q-values, this is relatively easy. The new policy is simply the following:

$$\pi(s,a) = \arg\max_a \{Q(s,a)\}. \tag{9.6}$$

This means that in the new policy, in state s, the action a leading to the highest estimated value ($Q(s, a)$) is chosen. For example, starting from the value system in figure 9.2d (and ignoring for simplicity the distinction between the V and Q values), the policy in state 1 would become "Go right" because that is the highest value (i.e., 8.9) that one can reap starting from state 1. Starting from that state (state 2), the best option is to go right again (4.5 is the best that one can get starting from state 2), and so on. Doing this for all states yields a better policy than the original random one.

If one has estimated V- instead of Q-values, finding a new updated policy can be a bit more difficult and may require several updating steps. However, as previously mentioned, in simple cases like the gridworld discussed in this chapter, the two procedures (using Q or using V for finding a better policy) are equivalent. I refer again to Sutton and Barto (2018) for further information.

Policy Iteration

In the previous example, if the new policy is determined, the value $V_\pi(s)$ [or $Q_\pi(s,a)$] of this new policy can again be estimated. After this estimation step, a new policy can again be computed, and so on. Repeated alternation between the two steps is called *policy iteration*.[2] Obviously, such an alternating process is feasible only for agents with huge amounts of time and memory on their hands. Taking the game of chess again as an example, it would require first estimating all 10^{45} $V(s)$ values, then updating the policy based on those 10^{45} $V(s)$ values, followed again by estimating all 10^{45} $V(s)$ values, and so on.

A more ecologically plausible approach would be to immediately update the policy after a few value estimation updates (or maybe even just one). Such an approach is called *generalized policy iteration* (GPI), and this is the approach most typically taken. Fortunately, it turns out that GPI also leads to an optimal policy.

Exploration and Exploitation in Reinforcement Learning

The policy as described thus far always chooses the best action according to the V or Q value. One says that the policy is *greedy* with respect to Q. However, to keep exploring uncharted (potentially better!) territories, we must include some random component in the greedy policy. For example, suppose that two sources, A and B, lead to rewards with probabilities .40 and .99, respectively. Obviously, it is best to always sample from source B in this case. But it is also very possible that one initially receives a reward from source A. If one would then (greedily) always follow source A based on this initial success [again, as equation (9.6) suggests doing], then the resulting policy will clearly be suboptimal. Thus, RL always requires a sufficient amount of exploration. At the same time, it also requires that one knows when to stop exploring. Indeed, if one knows that B is the better source, it should be aggressively exploited: in this case, it is suboptimal to ever sample A again. Again, this balance is the exploration-exploitation dilemma already mentioned in the other RL in chapter 8. This problem is hard to solve because the reward statistics of the environment depend on the statistics of the environment, which can be known only by exploring.

Different approaches have been developed to deal with this dilemma. One simple approach is to sample a random action with some probability ε (e.g., $\varepsilon = 0.05$); and thus the best action a is chosen with probability $1 - \varepsilon$. In this case, one calls π an ε-greedy policy with respect to Q. If ε is very high (say, $\varepsilon = .95$), the process is very exploratory: usually, a random action is taken. This helps one to learn the values of states and actions, but it doesn't follow the implemented policy very well. In contrast, if ε is very low (say, $\varepsilon = .05$), then the implemented policy is very well followed, but the consequences of actions that are assigned a very low value according to the policy will not be well estimated because they are never chosen. So in this case, one exploits one's knowledge but possibly does not explore enough to learn a better policy. It's best to choose an ε that balances between these two extremes. The ε-greedy approach is easy to understand and implement, but it does not well capture the fact that one should start with sufficient exploration and gradually shift toward exploitation for efficient policy iteration. The following approaches perform better in this respect.

The second approach is called the *softmax sampling procedure* (or simply *softmax*). This is the approach most often used in cognitive neuroscience RL applications (i.e., when RL is used to model human behavior). Here, the agent samples each action with a probability proportional to its exponentiated value. Thus, actions with higher estimated value are sampled more often (exploitation), but other actions are sometimes sampled as well (exploration). Specifically, one samples action i according to the following (in simplified Q-notation):

$$\text{Pr (action } i) = \frac{\exp(\gamma Q_i)}{\sum_k \exp(\gamma Q_k)}. \tag{9.7}$$

Here, I again use the inverse-temperature parameter γ. As in earlier models (e.g., in chapter 8), this parameter controls the amount of noise in the system. In the current application, this amounts to the balance between exploration and exploitation. In equation (9.7), if γ is very high, it would entail quasi-deterministic responding; lower values allow more exploration (see also the example worked out in chapter 6). Thus, by choosing the exploration parameter γ appropriately, one can attempt to find a balance between exploration and exploitation.

A third approach for dealing with the exploration-exploitation dilemma is called *optimistic initial values*. Here, each estimate starts with a high (i.e., optimistic) value in

Reinforcement Learning: The Markov Decision Process Approach

order to guarantee that all actions are initially sampled frequently. The unfavorable options are then gradually reduced in value. One algorithm that can be considered as an instance of optimistic initial values is the *upper confidence bound (UCB) method*. Here, one calculates not only the value, but also the uncertainty (inverse confidence; denoted U) of each action, and also deterministically chooses the option that has the highest quantity $Q + U$. Thus, if an action has high value (high Q) or high uncertainty (high U), that option is chosen more often. The Q term implements the exploitation part, while the U term implements exploration. Although softmax sampling is far more popular than UCB in cognitive neuroscience, recent computational model comparisons demonstrate that UCB may be more plausible in human learning and decision making (Stojic et al., 2020).

Finally, softmax and UCB can be combined by changing the argument of equation (9.7) to $\gamma(Q + \tau U)$. Here, parameter γ can be considered an undirected form of exploration; it simply adds more noise to each option. In contrast, the parameter τ reflects directed exploration: it lets the agent specifically sample those options about which less is known (i.e., from those options with higher uncertainty U). Computational model fitting on developmental data has demonstrated that children (relative to adults) exhibit more directed exploration (higher estimated τ) but equivalent levels of undirected exploration (Schulz et al., 2019; Wu et al., 2018). Also context may determine the parameter settings: for example, one study found that adults showed (specifically) lower directed exploration (lower estimated τ) in a high working memory load relative to a low load context (Cogliati-Dezza et al., 2019).

Applications

Temporal Difference Learning in the Primate Brain

One reason for the popularity of RL in recent decades is the discovery that dopamine neurons in the monkey midbrain seem to implement a learning rule that resembles temporal-difference learning (Ljungberg, et al., 1997; Seymour et al., 2004). That is, the authors observed that dopamine neurons in the midbrain respond only to *predictors of* reward. Specifically, in the initial stages of learning (when reward could not be predicted yet), these neurons respond to actual reward onset (e.g., food). But after learning, the same neurons respond only to cues that predict the reward, no longer to the reward itself (see figure 9.4). The temporal difference model also exhibits such a response pattern (Montague et al., 1996). Indeed, the temporal difference learning rule considers both reward itself (i.e., r_{t+1}), as well as *prediction* of reward [e.g., $V(s_{t+1}]$ as valid updates for calculating value [see equation (9.3)]. Later work has shown that the same dopamine neurons also encode other quantities, including the advance knowledge (and its prediction errors) of what reward the animal will receive (independent of the actual reward; see Bromberg-Martin & Hikosaka, 2009).

Later functional magnetic resonance imaging (fMRI) studies also found temporal difference–like signatures in the human basal ganglia (Seymour et al., 2004) and

**Do dopamine neurons report an error
in the prediction of reward?**

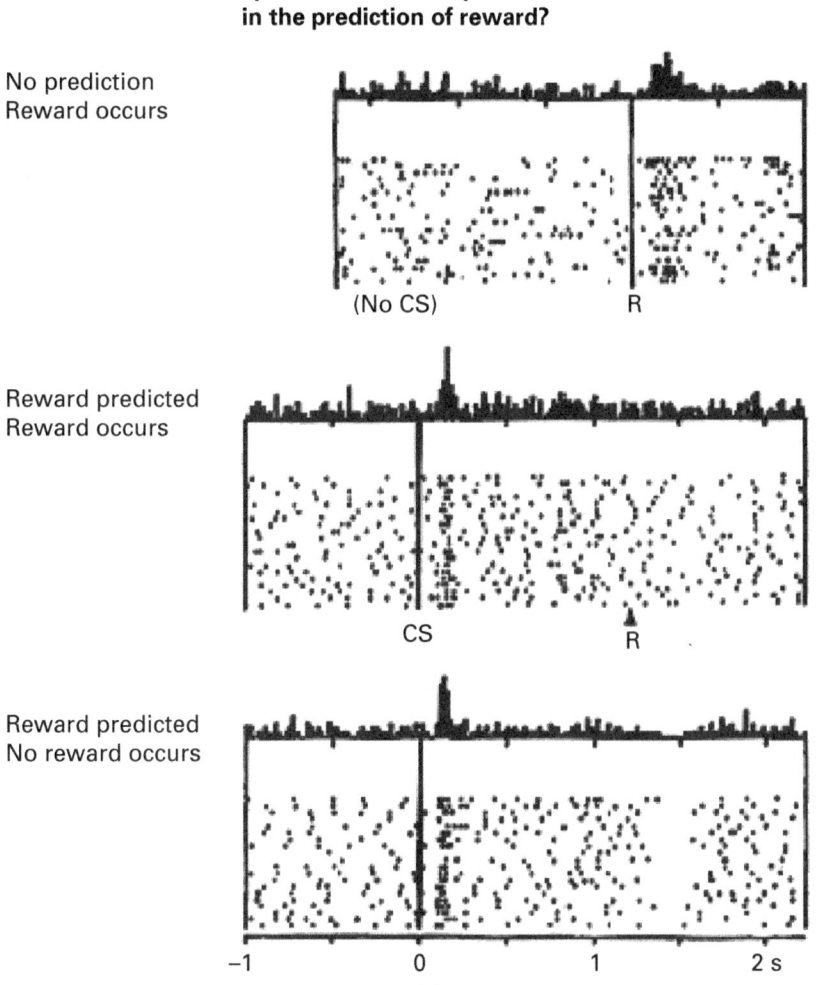

Figure 9.4
Temporal difference–like signatures of dopamine neurons. This figure was reproduced with permission from Schultz et al. (1997).

midbrain (D'Ardenne et al., 2008). For example, D'Ardenne et al. (2008) presented cues that predicted a liquid reward (juice) to thirsty human subjects. Some cues predicted juice; some did not. In some cases, the prediction of the cue was violated (e.g., it predicted a reward but none actually followed the cue). In this way, they could investigate the response of the midbrain nuclei that exhibited a temporal-difference signature in monkey. The authors observed the largest BOLD response when an unexpected juice was given; an intermediate response for expected juice; and the weakest response (below baseline) when juice was expected but not provided. This suggests that the human midbrain also encodes temporal differences.

The role of temporal differences has also been studied extensively using electroen-cephalography (EEG) in the human brain, starting from Holroyd and Coles (2002). These authors studied *feedback-related negativity* (FRN), a negative-going event-related potential wave, locked at feedback and measured at central electrodes at the scalp. The authors observed an increased negativity at error feedback, but only in trials where the error could not already have been predicted at response onset. In particular, for stimuli followed by random feedback, the participant cannot predict anything at response onset: she has to wait for feedback, and an FRN is observed here. Instead, in stimuli that lead to predictable feedback, one can already predict the feedback (right or wrong) at response onset. In such stimuli, errors tend to be rare and surprising, and so the electro-physiological negativity can already be observed at response onset, but not at feedback. This response-locked wave is typically called *error-related negativity* (ERN). This finding provides a striking parallel to the original data (Schultz et al., 1997) and has led to a large body of literature on prediction error processing using EEG and how such pro-cessing is impaired in clinical disorders (Gehring & Willoughby, 2002; Hajcak, 2015; Holroyd & Umemoto, 2016; Oliveira et al., 2007; Sambrook & Goslin, 2015).

Probabilistic Decision Making

Consider again the decision-making contexts from earlier chapters. Subjects learn the values of a number of stimuli (e.g., N-armed bandits), each of which provides a probabi-listic reward. In an MDP RL framework, subjects approach this problem by computing the value of each stimulus and then probabilistically choosing the stimulus that maxi-mizes value on each step. This approach can be considered as an instance of GPI. This is indeed the approach discussed in chapter 6; but now we see that this is a principled (i.e., approximating optimal) approach to solving this problem.

Because we simply choose one stimulus (out of N) for each trial, the notions of stimu-lus and action are indistinguishable and V and Q notations become interchangeable. We can thus update on each trial the following quantity using the Rescorla-Wagner rule:

$$V(s_t) \leftarrow V(s_t) + \beta\left(r_{t+1} - V_t(s)\right)$$

for the stimulus that was chosen by the subject; and V-values remain unchanged for the other stimuli.

The subject chooses a stimulus on each trial according to the following softmax rule:

$$\Pr(s_i) = \frac{\exp(\gamma V_i)}{\sum_j \exp(\gamma V_j)}. \tag{9.8}$$

Equation (9.8) ensures that one generally follows the policy consistent with the V-values; but at the same time, it leaves room for exploration.

This model has been used extensively in recent years as a vehicle to understand healthy and impaired decision making. Earlier in this book, I mentioned Hauser et al. (2014), who demonstrated increased exploration (decreased γ) in an attention defi-cit hyperactivity disorder (ADHD) population. A whole new field, called *computational psychiatry*, investigates how specific computational model parameters are impaired in various clinical conditions (such as ADHD); I refer readers to Maia and Frank (2011) for a general overview.

Another related approach is to compare model parameters across subjects with different alleles of specific genes. In a modified version of this model, Frank et al. (2009) distinguished between a learning rate for positive prediction errors (β_+) and a learning rate for negative prediction errors (β_-). They compared carriers of the T/T allele versus carriers of a C allele of the *DARPP32* gene, which controls striatal dopamine function. They found that the T/T carriers had an increased β_+ versus β_- learning rate relative to the C carriers. This is consistent with the idea that striatal dopamine is necessary for learning from reward. The *COMT* gene, which controls prefrontal dopamine, did not modulate β_+ versus β_-, but it did influence how subjects explored in the task used by these authors. In another study investigating how genetic information modulates behavior, den Ouden et al. (2013) demonstrated dissociable effects of serotonin and dopamine genes on parameters of the decision-making model.

The RL decision-making framework has been extensively used in a model-based fMRI framework. Here, computational model parameters are first estimated on a subject-specific basis, typically based only on behavioral data (see chapter 6). From these estimates and the model architecture, trial-by-trial regressors can be generated, which can then be included in an fMRI 1st-level analysis (see chapter 7; and Daw et al., 2011). Thus, brain areas that code for specific aspects of the decision-making models (e.g., prediction, prediction error) can be identified. As mentioned before, fitting the model parameters is not always needed; a regressor with optimal parameters and a regressor with random parameters will be very highly correlated, so the two regressors will typically covary with the same brain area (Wilson & Niv, 2015).

Although less common, the same model-based RL approach can be applied to generate regressors for other measurement modalities like pupillometry (Cheadle et al., 2014), magneto-encephalography (Hunt et al., 2015), and EEG (Collins & Frank, 2018; Wyart et al., 2012).

RL for Abstract Actions

A recent trend in cognitive neuroscience is to conceptualize the learning of abstract actions via RL. In this way, the computationally strong framework of RL can be applied to cognitive domains where it is not obviously applicable, such as attention, cognitive control, language, and working memory (Abrahamse et al., 2016). For example, we can all keep relevant information in working memory (at least some of the time). But how do we do it? How do we know which information to keep in working memory and what information to take out; and how do we know when retained information is needed for action? We must be careful in our theorizing not to posit a homunculus that "just knows" when and what information is relevant. Avoiding the homunculus trap can be done by considering "keeping information in working memory," "retrieving information from working memory," and related cognitive processes as actions whose

Reinforcement Learning: The Markov Decision Process Approach

Q-value can be learned via RL principles. Appropriate working memory actions can then be taken by choosing the action with the highest *Q*-value, just as one would in ordinary RL. Such movements in and out of working memory are called *gating*; one opens and closes, as it were, appropriate gates for information flow in and out of working memory.

One of the very first papers that described how abstract actions such as gating can be learned was Hochreiter and Schmidhuber (1997). Their approach was to consider these "gates" as parameters than can be updated via error minimization, as in the approach described in chapters 4 and 5. The first formulation of working memory gating in an RL framework dates back to O'Reilly and Frank (2006). Later fMRI work found neural evidence for the principles proposed in this model (Chatham et al., 2014).

Several other abstract actions have been modeled as subject to RL principles. Examples include error and conflict processing (Silvetti et al., 2014), effort investment (Niv et al., 2007; Shenhav et al., 2013; Silvetti et al,, 2018; Verguts et al., 2015), language production (Kriete et al., 2013), synchronizing cognitive processing modules (Verbeke & Verguts, 2019), and learning to learn (Wang et al., 2018).

Finally, another abstract action could be to choose between acting in a Model-free manner (in the Rescorla-Wagner sense, without using a Model of the world) versus a Model-based manner (i.e., using a Model of the world, such as state transition probabilities; Daw et al., 2011). Recent work has investigated to what extent human behavior is Model-free or Model-based, and under which conditions. For example, one study found that providing extra incentives makes human performance shift toward more Model-based behavior (Kool et al., 2017). Factors that shift human behavior to be more Model-free are working memory load (Otto et al., 2013a) and stress (Otto et al., 2013b).

Combining Gradient-Ascent and MDP Approaches

I have discussed gradient and MDP approaches toward RL. Which one is best? Gradient-based approaches to RL, discussed in chapter 8, are closely related to supervised-learning gradient methods, so they have similar pros and cons. On the con side, they can get stuck in local optima. On the pro side, gradient-based approaches allow a very natural parametrization of the input space, which is necessary in every realistic problem. Indeed, as mentioned before, one cannot learn chess by considering all its 10^{45} possible states. Strong generalization is needed between states, so one also can act adaptively in states that were not yet visited. Thus, the input space must be parameterized in some way. For example, one could parameterize the input space by $2 \times 16 + 1$ units for each board position (i.e., one unit for each of the 2×16 chess pieces, $+1$ for the empty position). As there are 64 squares on a chessboard, this approach would require $64 \times (2 \times 16 + 1)$ input units. A large number, but it is dwarfed by the original 10^{45}, which makes the chess problem tractable.

The pros and cons of the MDP approach are symmetric: they guarantee arriving in the global optimum, but because each state and action is treated separately, they do not scale up very well to real-world problems. Because of this trade-off between pros and cons, current AI algorithms combine aspects of both approaches.

One of the first models to combine both was the TD-Gammon program (Tesauro, 1995). This model's goal was to compute the value (probability of winning) for each possible state of the game of backgammon. As there are obviously too many states to represent explicitly, states were approximated by a multilayer neural network that projected to an output layer that represented the value of the state. All weights were trained by a variant of temporal-difference learning that backpropagated the value from the output layer to the deeper layers, just as one backpropagates error to the deeper layers in the standard backpropagation algorithm (see chapter 5). After playing thousands of games, the model eventually became an expert backgammon player.

A more recent example is the deep Q-learning program of Mnih et al. (2015) by the Google DeepMind group. This program learned to play forty Atari games. What was especially noteworthy is that all the games used the (almost) raw pixel input from the screen; and all games used the same network (including learning rate). Thus, the model was given the same very severe restrictions that confront a human learner, who also must process raw pixel input and (presumably) only has one brain. So how did it do that?

First, input was transformed via several layers to a Q-value output layer. This layer was composed of several components, with each component coding for the possible responses of one of the games. Each component attempted to measure the quantities $Q(s, a)$, the values of actions a in a given state s.

We can then consider choosing as our target for $Q(s, a)$ the reward at $t + 1$ (r_{t+1}) plus the best possible option on that time step [as in the Q-learning algorithm; see equation (9.5)]. Further, we again discount rewards obtained at later time steps t by the factor η. Formally, this would lead to the following goal function:

$$\sum_t \left(r_{t+1} + \eta \max_a \{Q(s_{t+1}, a)\} - Q(s_t, a_t) \right)^2.$$

In analogy to the gradient descent approach in supervised learning, we can consider the following algorithm:

$$\Delta w_{ij} = \beta (r_{t+1} + \eta \max_{a'} \{Q(s_{t+1}, a')\} - Q(s_t, a)) \frac{\partial Q(s_t, a)}{\partial w_{ij}}. \tag{9.9}$$

The gradient $\frac{\partial Q(s_t, a)}{\partial w_{ij}}$ can be propagated through several deeper layers, again as one would do with error (chapter 5) or with reward (chapter 8) in standard backpropagation. Thus, the updates Δw_{ij} for units residing in deeper layers can be computed. The resulting algorithm is called *deep Q-learning*.

I have ignored a number of technical aspects in the description of the deep-Q algorithm, but one deserves some mention, as it is also of cognitive interest. Specifically, the authors wanted to make sure that the algorithm has sufficient exposure to all relevant corners of the (state, action) space. Moreover, it is well known that one approach to avoiding catastrophic interference (see chapter 5) is to mix older with novel information during learning (McClelland et al., 1995). For these two reasons, the model was not trained online; instead, all experienced instances of $(s_t, a_t, s_{t+1}, a_{t+1}, r_{t+1})$ were stored, and the model learned via random sampling of these 5-tuples.

Note that in equation (9.9), such a 5-tuple is indeed sufficient to carry out one weight update step. This offline learning technique is called *experience replay*; it was directly inspired by theories of animal memory, according to which information acquired during the day is replayed during sleep for storage and integration with older information in memory, likely in the hippocampus (Hassabis et al., 2017; McClelland et al., 1995). However, which events must be replayed? In their deep-Q model, Mnih et al. (2015) replayed randomly sampled episodes, but a recent alternative approach implements *prioritized* experience replay: here, events with a large prediction error are preferentially replayed. Recent models of hippocampus implement prioritized experience replay (Mattar & Daw, 2018), and recent work also documents an important role for reward prediction error in human episodic memory (Ergo et al., 2020; Rouhani et al., 2018).

Reinforcement Learning for Human Cognition?

Given their current success and attention, a relevant question is to what extent the recent generation of AI (multilayer and RL) algorithms, as discussed mainly here and in chapter 5, are plausible as models for human cognition. Some argue that they are not; they are *idiots savants* who can profit from extreme quantities of data and computational power. Their performance is unlike how humans (with much lower computational power and data available) solve such tasks (Lake et al., 2017; Marcus, 2018). For example, humans can quickly change according to (partially) novel rules; they could immediately (and presumably, not too badly) play chess when the pawns were suddenly granted extra movement opportunities. Such a change would require significant relearning in computer algorithms. Moreover, most of these models can solve only a limited number of tasks (often only one), whereas humans can switch smoothly among a rich array of tasks. Scaling up AI (and indeed, cognitive) models toward full human intelligence is currently an active area of research. As anticipated at the start of chapter 8, further progress will likely require an integration of RL principles with Models of the world. This trend is already ongoing: the AlphaGo program that beat the Go world champion (Silver et al., 2016) combined RL principles with searching a decision tree in order to prepare each move; a decision tree is just one instantiation of a world Model.

More than just Models, humans construct causal Models of the world. They understand that dark skies will cause rain, but either staying inside or bringing an umbrella can prevent one from becoming wet. Implementing such knowledge of how one event causes or prevents another is still challenging for AI. However, finding out how humans and other primates solve complex tasks is likely a fruitful field where cognitive neuroscientists can continue to inspire this domain (Hassabis et al., 2017).

Open AI Gym

So you want to try RL algorithms but don't know where to start. If your focus is on model analysis and you have data for a task that can be conceptualized as a contextual *N*-armed bandit, a popular choice is to fit the learning and decision-making model (e.g., code in the chapter 6 GitHub folder). If your focus is on model building, you can download the Python package Open AI Gym (from the company OpenAI), which contains a number of environments with problems to solve, ranging from very basic to more challenging (such as making a robot walk). Here, you can construct any model (e.g., using TensorFlow) to interact with the environment. You find some examples in the chapter 9 GitHub folder.

To sum up, the MDP RL framework implements a two-step approach to reward maximization, alternating between value estimation and policy updating. In recent decades, the framework has been a fruitful domain of research for neurophysiology, cognitive science, and AI, with several interactions among all of them. Progress has been achieved by gradually moving RL outside its comfort zone of stimulus-action learning and toward building Models of the world and using them for optimizing performance. Further progress will likely require continuing this trend.

10 Unsupervised Learning

I did it my way.
—Paul Anka

In earlier chapters on learning (3-5, 8-9), there was always an external teacher telling the organism what to do. In the reinforcement learning (RL) case, this external teacher was quite minimalist in communication ("you're doing well" or "you're doing badly"), but at least it provided something. In this chapter, I consider the case where there is *no* feedback from the environment; this approach is called *unsupervised learning*.

So what can an agent learn in this case? It turns out that plenty can be learned from simply observing the environment. One can learn about the *structure* of the environment (more formally, the statistical distribution of variables appearing in the environment). In the cats-and-dogs example, one can learn that having one's picture on Facebook and biting visitors are negatively correlated, merely by observing the environment (and thus receiving no feedback). This is a very fortunate situation because in real life, we rarely receive explicit rewards (chapters 8 and 9) or explicit feedback on what we should have done (chapters 3–5). Thus, unsupervised learning is undoubtedly of crucial importance in human and animal learning. I discuss a few of the most popular unsupervised learning rules.

Unsupervised Hebbian Learning

Suppose that we have a network with just a single output unit, as in figure 10.1a. The input flowing into this output unit is

$$y = \sum_{j=1}^{J} w_j x_j. \tag{10.1}$$

In contrast to the Hebbian learning rule from chapter 3, there is no external feedback; in other words, we now have no supervisor telling us what the output should be.

Instead, one could consider it a worthy goal to compress as much information as possible about the original (*J*-dimensional) input with this single output unit. For example, if the *Facebook*, *Four Feet*, and *Biting* variables are perfectly correlated, then this information can be compressed in a single variable.

In statistics, principal component analysis does exactly that. Specifically, suppose we have a collection of *J* variables (each of which is mean-centered first: see figure 10.1b for an example with *J* = 2). The principal component is then the variable that extracts the one-dimensional (1D) commonality among the original *J* variables as faithfully as possible.[1] Consider again figure 10.1b: one can project each data point into a line. Then, one can calculate the reconstruction error for each data point; the length of the black arrow in figure 10.1b is the reconstruction error for a single data point. The line that yields the lowest reconstruction error averaged across all data points, is the principal component. This principal component will run through the data points, as illustrated in Figure 10.1b. The coordinates on this line constitute the principal component variable. It is clearly 1D because after projection, all data points fall exactly on the (1D) line.

In statistics, one applies standard linear-algebra algorithms to compute this projection onto this line, and thus the principal component *y*. If we consider optimization from a model's perspective, then such linear-algebra algorithms cannot be applied because they are infeasible to perform online; recall the discussion in chapter 6 contrasting optimization from a model's versus a modeler's perspective. However, it turns out that principal component extraction can also be iteratively implemented. In particular, the following unsupervised Hebbian learning rule iteratively extracts the principal component:

$$\Delta w_j = \beta x_j y. \tag{10.2}$$

So if one applies the learning rule in equation (10.2), one ends up with an output unit that extracts the first principal component of the data. To illustrate, figure 10.1b shows data points that were presented on a trial-by-trial basis to a model with two input units, and hence two-dimensional (2D) input. The line represents the weight vector after learning via equation (10.2).

The fact that the weight vector aligns along the principal component (direction of minimal reconstruction error) is interesting and intuitive, but unfortunately, this learning rule has a major stability problem. In particular, just as we saw in the supervised Hebbian learning rule in chapter 3, the weights *w* tend to drift off to infinity (even though the principal component variable, *y*, does not). This makes the solution unstable; small perturbations in input can cause large weight changes. This is not what we want.

To address this instability problem, different corrections methods for the unsupervised Hebbian learning rule have been proposed. One elegant solution is to subtract an appropriate term from the weight update [equation (10.2)]. For example, in Oja's learning rule (Oja, 1982), the weight update is

$$\Delta w_j = \beta \left(x_j y - y^2 w_j \right). \tag{10.3}$$

The intuition behind this rule is that if weight *w* or output *y* becomes large, then the baseline $y^2 w$ for a positive Δw becomes large as well. As a result, the weight change Δw

Unsupervised Learning   155

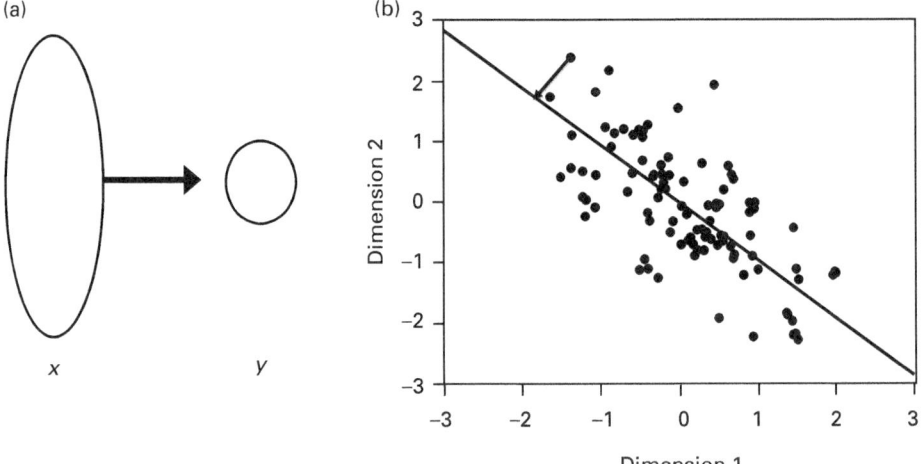

Figure 10.1
(a) Two-layer unsupervised model architecture with one output unit. (b) Data and Hebbian weights (straight line; aligned with the principal component). The perpendicular projection into the principal component is illustrated for one data point. The black arrow is the residual of this projection; its length is the reconstruction error discussed in the text.

is reduced, and thus also weight w itself will be reduced (or at least, grow less quickly). In this case, the variable y also tends toward the principal component of the correlation matrix, but the weight vector $\mathbf{w}$ tends toward a length of 1. In this way, Oja's rule constitutes a stable variant of unsupervised Hebbian learning. In practice, however, this precaution is often still not sufficient to keep the weights within reasonable bounds (as you can find out by trying the code online). A drastic but effective solution is to explicitly normalize the weights on each trial. In whatever way it is implemented, it is clear that weights must be normalized in some way, and there is indeed evidence that synapses do so (Bourne & Harris, 2011).

The principal component is also called the *first principal component*; there are also other principal components—the second principal component extracts the most "remaining" variance after accounting for the first principal component, the third extracts the most variance after accounting for the first and and second, and so on. Intuitively, one could consider adding extra output units with activation $y_2, y_3, \ldots$, and a learning rule analogous to Oja's learning rule [equation (10.3)]. But it is not the case that the respective activations $y_2, y_3, \ldots$, will tend toward the second, third, $\ldots$ principal components. This problem can be solved by subtracting an appropriate term on the right side of equation (10.2) (Sanger, 1989). There are several other corrections and generalizations to the basic unsupervised Hebbian learning rule. I refer readers to specialized literature for further information (e.g., Cooper & Bear, 2012; Hinton, 1989).

One interesting application of unsupervised Hebbian learning in cognitive neuroscience is the work of Linsker (1986a, 1986b, 1986c). He constructed a multilayer network receiving random visual inputs (at the input layer); each pair of adjacent layers was trained with an unsupervised Hebbian learning algorithm. He observed that layers in such a system spontaneously develop the properties observed in biological cells in the visual system, as originally described in the seminal neurophysiological studies of Hubel and Wiesel (1968). Specifically, in the first few layers, Linsker observed a spontaneous emergence of spatial-opponent cells—that is, with either an on-center activation profile (i.e., active when light falls into the center but not the surround of its receptive field), or an off-center activation profile (i.e., active when light falls into the surround but not the center of its receptive field) (Linsker, 1986c). In subsequent layers, he observed an emergence of orientation-selective cells—that is, cells that were sensitive to either bars or edges with a specific orientation in a specific part of the visual field (Linsker, 1986b). Finally, cells in the final layers of his multilayer network spontaneously organized into orientation columns (i.e., patches of adjacent cells with the same orientation selectivity), again as observed in primate visual systems (Linsker, 1986a).

Competitive Learning

A second type of unsupervised learning is competitive learning. Here, output units enter a competition in order to be able to represent the current input vector. Consider figure 10.2a.[2] It represents a (very small) competitive network, containing two input units and two output units. Let's assume that the input variables are continuous; then we can represent all possible inputs in a 2D plane as in figure 10.2b. The two input dimensions could represent the ways (variables) in which cats and dogs differ among one another. Each input *pattern* is represented by a dot in figure 10.2b. Each output *unit* is represented as a cross in figure 10.2b. Note that it has the same dimensionality as an input pattern because it has two weights attached to it. More generally, if the input is J-dimensional (i.e., there are J input units), then each output unit has J weights attached to it.

Cats could represent one of the two large dot clouds in figure 10.2b; dogs could represent the other. The goal of competitive learning is now to represent this (here, bimodal) input distribution as faithfully as possible. Thus, we will attempt to place the output units (the crosses) in the input distribution in such a way that the input distribution is covered as accurately as possible. In this case, this would correspond to placing one of the crosses in one of the large dot clouds (near its center of gravity); and the other cross in the other large dot cloud. Note that the covering of the input distribution is very rough because we have just two output units available. If we had a third output unit (cross) available, we could also place that unit in the smaller (atypical dogs? wolves?) cloud at the top of figure 10.2b. Generally, the more output units there are, the more faithful is our representation of the input distribution.

How do we achieve that? Suppose that one of the input patterns is presented. Then each of the two output units i computes its activation in the familiar manner:

$$y_i = \sum_{j=1}^{J} w_{ij} x_j.$$

(10.4)

Unsupervised Learning 157

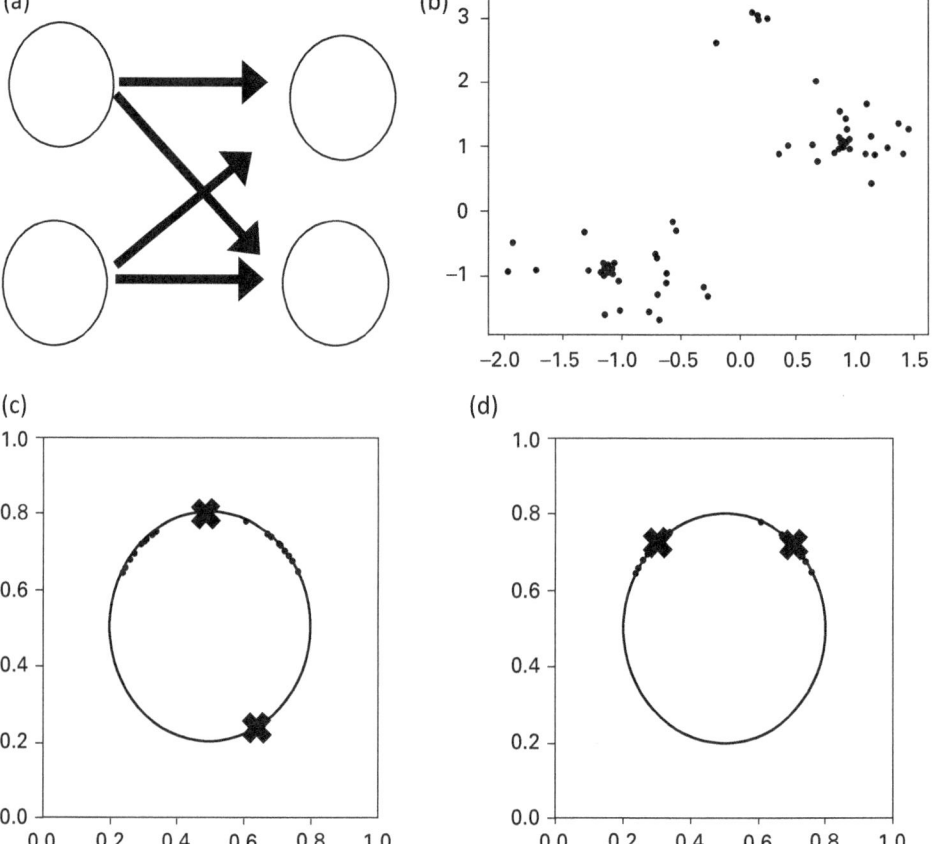

Figure 10.2
Competitive learning. (a) Architecture. (b) Data and output vectors (unnormalized, after learning).
(c) Data and output vectors (normalized, before training). (d) Data and output vectors (normalized, after training).

Crucially, however, let's now suppose that all input vectors are normalized (with length 1); that is,

$$\|\mathbf{x}\| \equiv \sqrt{\sum_{j=1}^{J} x_j^2} = 1,$$

and also the weight vectors $\mathbf{w}$ attached to each output unit are of length 1: $\|\mathbf{w}\| = 1$.

Then, supposing for convenience that the input space is 2D, we can represent both input vectors and weight vectors on a circle, as in figure 10.2c. Obviously, in J dimensions, this becomes a J-dimensional sphere, but this is hard to represent on a 2D screen or paper. With these length constraints, maximizing the sum [equation (10.4)] can be done by making $\mathbf{w}$ as close as possible to $\mathbf{x}$. This follows from the mathematics of dot products, but I leave the formal argument aside (but see exercise 10.3). For now, note that because the sum [equation (10.4)] is maximal when $\mathbf{w}$ is as close as possible to $\mathbf{x}$, this sum amounts to a measure of the inverse distance between $\mathbf{w}$ and $\mathbf{x}$.

The competitive part consists of the fact that the output unit with the maximal sum (smallest distance to the input pattern) is then allowed to be active ($y_{\text{win}} = 1$), whereas all other output units are inactivated ($y = 0$). This winning unit is then allowed to change its weights with the following equation:

$$\Delta w_{win,j} = \beta(x_j - w_{win,j}). \tag{10.5}$$

Intuitively, equation (10.5) attempts to balance the weights attached to the winning output unit such that it is "the middle" between all the input vectors $\mathbf{x}$ that this particular output unit represents. Thus, starting from a situation as in figure 10.2c, the weight configuration will evolve toward a situation as in figure 10.2d, where on average $x_j - w_{\text{win},j} \approx 0$.

Soft Competitive Learning

One application of competitive learning was reported in the numerical cognition literature (Verguts & Fias, 2004). These authors proposed a model in which output units were trained to respond to both nonsymbolic numbers (e.g., the number 4 represented by a cloud of four dots), and to symbolic numbers (e.g., the Arabic number 4) at input. The competitive learning rule was soft in the sense that not only the winning output unit was allowed to change its weights from input, but also other output units that were sufficiently close to the input pattern (this rule is called the neural gas algorithm; Fritzke, 1995). As a result, the authors found that specific output cells developed broad tuning curves for nonsymbolic input numbers. For example, a given output unit would respond strongly to four dots, somewhat less to three or five dots, even less to two or six dots, and so on. Furthermore, the same output units would respond to symbolic numbers, but with much more strongly peaked tuning curves; for example, an output cell that would prefer (respond most strongly to) the symbolic number 4, would respond much less strongly to 3 or 5, even less strongly to 2 or 6, and so on. This difference in tuning curves between symbolic versus nonsymbolic numbers is consistent with single-cell recordings performed in trained macaque monkeys and human functional magnetic resonance imaging (fMRI) data (Nieder et al., 2002; Piazza et al., 2007).

Exercises

10.1 Explain the role of the learning rate β in the competitive learning rule. Consider what happens in the extreme cases $\beta = 0$ and $\beta = 1$.

10.2 Demonstrate that the competitive learning rule can be considered a type of Hebbian learning.

10.3 Explain (look up, if necessary) why the length normalization constraint ensures that a dot product amounts to calculating a distance between two vectors. Would a sum constraint (e.g., $\Sigma x_j = 1$) also work?

Kohonen Learning

Another soft variant of competitive learning appears in the Kohonen network (Kohonen, 2001). A novelty compared to earlier models is that here, one defines a distance function

Unsupervised Learning   159

in the output space. In previous models, it made no sense to talk about which units were close together (or not) in the output space. This is different in the Kohonen network. Consider the model in figure 10.3a,[3] with three input units (x), and a 2D output map (y). Output units that are drawn close together in figure 10.3a are actually intended to be close to one another. For example, the two units in the upper-left corner are close together; a unit in the upper-left and another unit in the lower-right corner are far apart. The distance between two output units plays a role in the learning algorithm, as we will now see.

Each output unit i computes input in the usual linear way [e.g., as in equation (10.4)]. This is followed by a competition between the output units. Let's say that unit i_{max} wins the competition. All units are then allowed to change their weights as follows:

$$\Delta w_{ij} = \beta(n)e^{-\tau d(i,i_{max})}(x_j - w_{ij}), \tag{10.6}$$

where n is the trial number and $d(i, i_{max})$ is the distance between unit i and the winning unit, i_{max}. In a 2D map, the distance between 2D points a and b equals

$$d(a,b) = \sqrt{(a_1 - b_1)^2 + (a_2 - b_2)^2},$$

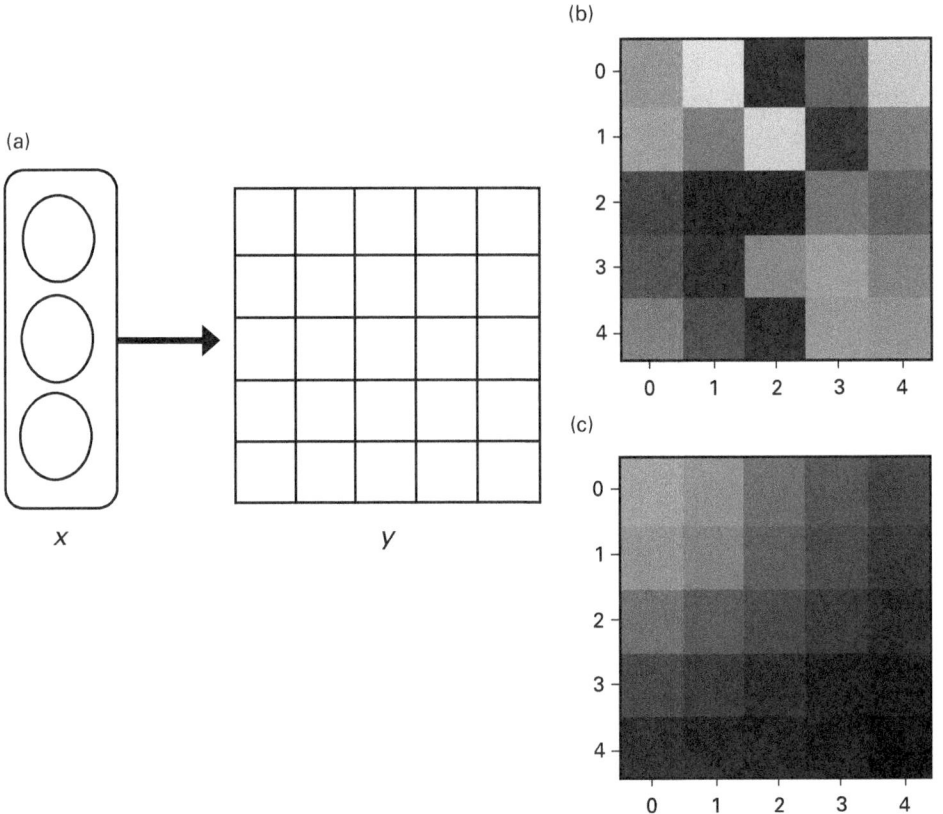

Figure 10.3

Kohonen network: (a) network architecture; (b) output representations before training; (c) output representations after training.

with a_i, b_i as the values along output map dimension $i = 1$ or 2. The parameter τ is another scaling parameter; it scales the effect of distance between two output units.

In general, we map a J-dimensional (input dimension) problem into a $I_1 \times I_2$–dimensional output field, with I_1 and I_2 the length and height, respectively, of the rectangle that constitutes the Kohonen map. Because output units that are close together in the output field undergo similar weight changes, they will tend to code for similar input vectors. As a result, one can inspect the 2D Kohonen map in order to investigate the input space structure. An example is shown in figure 10.3. As input, this network receives vectors representing color as in the red-green-blue (RGB) color system, with vectors $(1,0,0)$, $(0,1,0)$, and $(0,0,1)$ representing red, green, and blue, respectively. The initially random coding structure of the map is shown in figure 10.3b. After training, the coding of the map looks as in figure 10.3c. As can be seen, the structure of the input space is nicely represented in the 2D output map. In principle, also higher-dimensional output spaces are possible, but the number of weights increases exponentially with dimensionality. Further, the nice visualization property is lost in higher dimensions.

In equation (10.6), I made explicit that the learning rate β can be a function of time, particularly of trial number n. For example, one could stipulate that learning rate decreases as a power function of n, $\beta = \beta_0/n$. This principle says that changes should be large at first but then become more fine-grained as learning continues. Think of playing golf: at first, you want to hit the ball "in the ballpark" with a powerful club. Later, when you are closer to your goal, you want to use a finer, more precise one.[4] The same decreasing learning rate principle can be applied to models in previous chapters as well.

Most of the learning rules that I have discussed thus far optimize a well-specified goal function. In contrast, the Kohonen learning rule does not clearly optimize some goal function. The fact that a learning rule does not clearly optimize some function should not be surprising. If a learning rule is explicitly derived from gradient descent on a function (e.g., energy function), then obviously the learning rule attempts to optimize that function. In contrast, if the cognitive modeler develops a learning rule based on other (computational, biological, practical, aesthetical) considerations, rather than deriving it from the optimization of an energy function, then there is no guarantee that the learning rule optimizes some function. Some authors have changed the originally proposed Kohonen rule, such that the resulting update rule does optimize a goal function (Heskes, 1999).

Topographic Maps in the Human Brain

Kohonen maps are used in several branches of science to visualize high-dimensional input structures. For example, they are used to visualize the similarities and differences between genes (Nikkilä et al., 2002). For cognitive neuroscience, its relevance derives from the fact that also the brain has been proposed to contain 2D, or perhaps 1D or three-dimensional (3D), maps that represent high-dimensional input structure. Topographic maps are found throughout the brain. The classical maps are those for primary sensory and motor features. For example, both the occipital and parietal cortices contain a number of retinotopic maps, where the physical locations in visual space are approximately preserved. For example, when a sequence of concentric circles are presented, different parts of area V1 in a subject watching this sequence respond to different circle sizes. The same principle is found in several other posterior (occipital, parietal) cortical areas (Silver & Kastner, 2009). Similarly, auditory cortex cells are arranged in a frequency-dependent order: Different cells arranged in a 1D order respond to increasingly higher frequencies. Finally, primary somatosensory and motor cortex contain so-called homunculi that represent each part of the human body, with more cortical size devoted to more sensitive organs. In particular, the hands and mouth are much larger in the homunculi than their relative actual size in the body. These maps remain plastic in later life: for example, professional string players such as violinists have an increased cortical somatosensory representation of the left hand (the hand they use to manipulate the strings) (Elbert et al., 1995). In Kohonen map terminology, input patterns that receive more space in the output map can be represented with more precision.

In more recent years, maps for higher-order cognitive concepts also have been discovered, such as letter maps in the visual processing stream (Vinckier et al., 2007) and emotion maps (Koide-Majima et al., 2020). Computationally, several authors have shown that both low- and high-level maps can be developed by Kohonen-type networks (Li et al., 2004; Mayor & Plunkett, 2010).

Finally, recent authors have also suggested that working memory and its limitations can be understood as emergent properties from competitive-map interactions (Franconeri et al., 2013). They suggest that items in working memory must compete for being represented on physical maps in the parietofrontal cortex. This principle allows one to account for several findings from the working memory literature, such as the fact that there is an upper bound on the number of items that can be represented, which depends on the model's output dimensionality. Furthermore, the model predicts that when there are fewer items, each can be represented with more precision than if there are more items, which is again highly consistent with empirical data (Ma et al., 2014). Thus, this perspective may shed a novel light on the slots-versus-resources-versus-interference discussion on working memory that was discussed in chapter 6.

Auto-Encoders

One popular application of unsupervised learning is to auto-encoders. The goal of an auto-encoder network is to reconstruct the original input pattern as faithfully as possible. This is not as trivial as it may sound: in an auto-encoder, the original J-dimensional input is first mapped into a lower K–dimensional hidden layer (with $K < J$), after which it projects to a J-dimensional output layer, where the input layer must be reconstructed. In this case, we have a $J - K - J$ autoencoder, but a deeper auto-encoder (with more hidden layers) is possible as well. Because of the K-dimensional bottleneck in the hidden layer, the model is forced to compress the original information as well as possible. To achieve this, one could use the backpropagation algorithm from chapter 5, where on

162   Chapter 10

each trial, the target output pattern is exactly the same as the original input pattern. However, because no extra information is required beyond the input pattern (which is identical to the target pattern), unsupervised learning algorithms also can be used to train auto-encoders. Obviously, in the case of an auto-encoder, the boundary between supervised and unsupervised learning becomes a bit blurry. In the next section, I discuss a very popular unsupervised learning algorithm that is often used to train auto-encoders.

Exercises

10.4 Demonstrate that Kohonen learning is a generalization of competitive learning. Which parameter must be changed to go back to the original competitive learning algorithm from equation (10.5)?

10.5 In the Kohonen rule [equation (10.6)], what would happen if you accidentally switched w_{ij} and x_j? Check your answer by implementing the new "algorithm" in the Kohonen code.

Boltzmann Machines

In chapters 4 and 5, I discussed how to learn logical rules (e.g., OR, XOR) in unidirectional networks where information flows in one direction, from input to output units. Instead, one could construct an auto-encoder where the input and output units are exactly the same, and then apply the recurrent Hopfield model to implement such rules. Consider the simple model in figure 10.4a. If the patterns consistent with the OR rule—namely, $(0, 0, 0)$, $(1, 0, 1)$, $(0, 1, 1)$, and $(1, 1, 1)$—are attractors, as encoded by the weights, then the OR rule is effectively implemented in the model. Indeed, one can then clamp the two first units [e.g., to $(0, 1)$], and the network will autocomplete to the third unit [e.g., to (1)]. The attractors could be shaped (i.e., weights could be learned) via a Hebbian learning rule, which is indeed the learning rule typically used in the Hopfield model (see chapter 3). In this manner, an unsupervised network can be used for

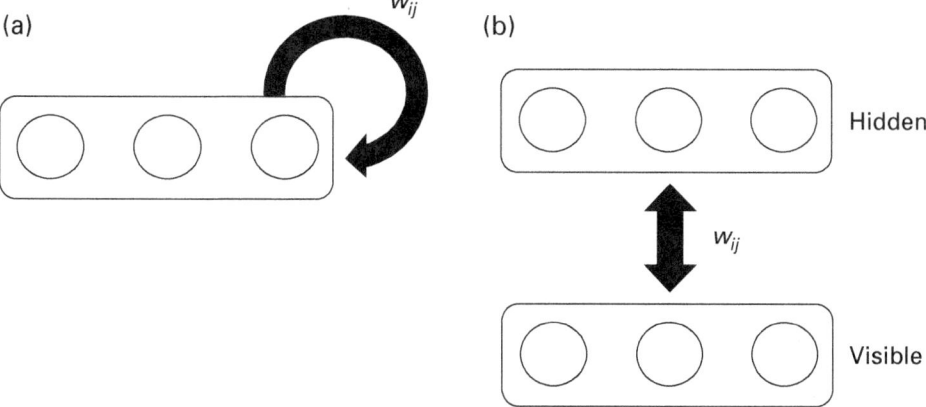

Figure 10.4
Boltzmann machine architectures: (a) Boltzmann model, no hidden units; (b) restricted Boltzmann model (with hidden units).

Unsupervised Learning   **163**

auto-encoding the patterns of a given logical rule. Unfortunately, the Hebbian learning rule is not very powerful, in the sense that it cannot be used to store many patterns (see also chapter 3). Furthermore, the Hebbian approach does not generalize to hidden units. Yet, hidden units are needed to implement logical rules such as XOR.

The Boltzmann machine addresses these problems. Like the Hopfield model, it is a recurrent model where the main goal is to autocomplete a pattern based on an incomplete version of that pattern. However, for the Boltzmann machine, there is a simple learning rule that also works with a network that contains hidden units (Ackley et al., 1985; Hinton & Sejnowski, 1983).

To explain the Boltzmann machine, one formulates the learning problem in a probabilistic framework. Consider an environment consisting of different states of the world, or simply states, each characterized by a vector $\mathbf{x} = (x_1, \ldots, x_I)$. For example, one state could be "a fluffy but nonaggressive cat"; if features are coded as $x_1 = $ fluffy, $x_2 = $ aggressive, and $x_3 = $ dog (instead of cat), then this state is represented as $(1,0,0)$. Consider now the statistical distribution across all possible states of the world, $\Pr(\mathbf{X} = \mathbf{x})$ (or $\Pr(\mathbf{x})$ for short). The Boltzmann machine defines an explicit statistical distribution over all possible states of the network $\Pr_w(\mathbf{x})$. The subscript w indicates that the probability depends on parameters w_{ij}. The goal of the model is now to mimic the empirical probabilities $\Pr(\mathbf{x})$. In particular, it will attempt to make $\Pr_w(\mathbf{x})$ as close as possible to $\Pr(\mathbf{x})$. So if a fluffy, nonaggressive cat appears in the real world with probability $1/10$ (i.e., $\Pr(\mathbf{x}) = 1/10$), that should also be true in the model-generated states ($\Pr_w(\mathbf{x})$).

So let's now consider how $\Pr_w(\mathbf{x})$ is defined. The probability that the model finds itself in state $\mathbf{x} = (x_1, \ldots, x_I)$ equals

$$\Pr_w(\mathbf{X} = \mathbf{x}) = \frac{\exp\left(-\frac{1}{T}\left(-\sum_j \sum_{i<j} w_{ij} x_i x_j\right)\right)}{Z}, \tag{10.7}$$

where T is a temperature parameter. Recall that I labeled γ as the inverse temperature before; so to keep notation consistent, I define $\gamma = 1/T$. Weight w_{ij} connects units i and j (in a symmetric manner: the weight from i to j is the same as the weight from j to i).

Equation (10.7) defines a statistical distribution called the *Boltzmann distribution* (hence the name of the model). The denominator Z is a normalization term to make sure that the summation across all possible patterns $\sum_{\mathbf{x}} \Pr_w(\mathbf{x})$ equals 1. Note that we can again define the term $-\sum_j \sum_{i<j} w_{ij} x_i x_j$ as the energy of the system (or machine, or model), as in the Hopfield model. The system attempts to minimize its energy, in the sense that low-energy states are assigned a higher probability. In other words, the system will often find itself in the low-energy states. If the system is very cold (i.e., temperature T is very low), this tendency to prefer low-energy states becomes more extreme. In contrast, if temperature T is high, the system will be "jumpier" and more often switch between high- and low-level energy states. Note that just like $\gamma > 0$, we also assume $T > 0$. If you like the physical analogy, you can think of T as measured in degrees Kelvin, where temperature is always positive.

The learning problem can now be formulated as follows: how do we make the distribution $\Pr_w(\mathbf{x})$ such that the empirically observed patterns $\mathbf{x}$ have high probability? If we can find such a distribution $\Pr_w(\mathbf{x})$, we can use the Boltzmann machine for several purposes, including the input-output mappings that were discussed in chapters 4 and 5. For example, in case we want to implement the OR rule, the patterns $(0,0,0)$, $(1,0,1)$, $(0,1,1)$, and $(1,1,1)$ should have high probability, and all "unobserved" patterns [e.g., $(0,0,1)$] should have zero (or at least very low) probability. This defines a probability distribution $\Pr(\mathbf{x})$ that we can approximate using $\Pr_w(\mathbf{x})$.

The distribution [equation (10.7)] is determined by the weights w of the network. The goal of the Boltzmann machine is to change these weights to mimic the desired probability function, $\Pr(\mathbf{x})$. This is a maximum-likelihood problem like the one we studied in chapter 6, except that now we consider the optimization from the model's perspective rather than from the modeler's perspective. Stated otherwise, the model is to supposed to perform the optimization, not the modeler.

Suppose that we actually observe pattern $\mathbf{x}$ with probability $\Pr(\mathbf{x})$. The desired probability $\Pr_w(\mathbf{x})$ generated by the model should be as close as possible to $\Pr(\mathbf{x})$. Recall that we already (although very briefly) introduced a maximum likelihood framework for network optimization in chapter 4, in the context of the cross-entropy error function. Analogous with equation (4.8), the log-likelihood of the data can be formulated as follows in this case:

$$\log L = \sum_{\mathbf{x}} \Pr(\mathbf{x}) \log\bigl(\Pr_w(\mathbf{x})\bigr). \tag{10.8}$$

Maximization of equation (10.8) is performed as a function of the parameters in $\Pr_w(\mathbf{x})$ [see equation (10.7)] to best match the data $\Pr(\mathbf{x})$. In addition to a log-likelihood that needs to be maximized, there are a number of other ways to think about equation (10.8). First, except for a constant, the quantity in equation (10.8) is equal to the "distance"[5] between two distributions, $\Pr(\mathbf{x})$ and $\Pr_w(\mathbf{x})$. The first distribution is imposed by the organism's environment, while the second is the model-predicted distribution. We thus can also say that we strive to make the model distribution $\Pr_w(\mathbf{x})$ as close as possible to the empirical distribution $\Pr(\mathbf{x})$. As a second alternative way to think about equation (10.8), consider that the quantity $-\log(\Pr_w(\mathbf{x}))$ measures the surprise of a given observation $\mathbf{x}$. Except for the minus sign, equation (10.8) is the average surprise (i.e., averaged across all possible states $\mathbf{x}$ of the organism's environment). Therefore, we can also say that we attempt to minimize average surprise by maximizing equation (10.8). In his Bayesian brain hypothesis, Karl Friston (2010) argues that minimizing a surprise measure similar to equation (10.8) is the ultimate goal of cognition (and indeed, perhaps of all life). I will explain this hypothesis in more detail in chapter 11, on Bayesian models.

As in earlier chapters, we want to find a suitable update rule that climbs the function $\log L$. Also as before, for that purpose, we take the partial derivative of $\log L$ toward each weight w_{ij} in order to implement gradient ascent. The update rule turns out to be the following (Ackley et al., 1985):

$$\Delta w_{ij} = \beta\bigl(p(i,j) - p_w(i,j)\bigr), \tag{10.9}$$

Unsupervised Learning   165

where β is a learning rate (in which temperature T is absorbed); $p(i,j)$ is the observed (environmental) probability that units i and j both take on the value of 1 (a function of all probabilities $\Pr(\mathbf{x})$); and $p_w(i,j)$ is the model-based probability that both units i and j take on the value of 1 (a function of all probabilities $\Pr_w(\mathbf{x})$).

Note that the update rule in equation (10.9) is again of the delta-rule variety: it compares a quantity imposed by the environment ($p(i,j)$, similar to the target t_i we had in chapter 4), with a network-generated quantity ($p_w(i,j)$, similar to the network-generated quantity y_i we had in chapter 4).

Table 10.1 shows a small-scale example of the probabilities with which a Boltzmann machine was trained (the model depicted in figure 10.4a, with a constant unit added). During training, I presented it all patterns consistent with logical AND, each with equal probability. For example, if the last unit is the output unit (and ignoring the constant unit for simplicity), then the allowed patterns according to the AND rule are $(0,0,0)$, $(1,0,0)$, $(0,1,0)$, and $(1,1,1)$. Each pattern was presented with probability ¼. In this data set, the probability that units 1 and 2 are both active is just the probability of pattern $(1,1,1)$, which is ¼. The only other possible data pattern where both units 1 and 2 are on (i.e., $x_1 = x_2 = 1$)—namely, $(1,1,0)$—has 0 probability in the AND rule.

After training with the learning rule [equation (10.9)], I clamped different pairs of units and observed the resulting probabilities of different patterns $\mathbf{x}$. These probabilities are shown in table 10.1. As can be seen, the model implements the logical AND. For example, if the first two units are clamped to 1 and 0, respectively, then the only high-probability pattern is $(1,0,0)$.

Although the update rule [equation (10.9)] looks simple enough, it requires two probabilities, each sampled from a different equilibrium (steady-state) distribution. For generating the model probabilities $p_w(i,j)$, one first can calculate the probabilities $\Pr_w(\mathbf{x})$ using equation (10.7) and then directly calculate $p_w(i,j)$ from them (see exercise 10.7). Alternatively, we could estimate the required probabilities $p_w(i,j)$ indirectly, by simulating data

Table 10.1

Boltzmann machine for implementing the AND rule

Clamped $\rightarrow$ Pattern $\mathbf{x}$	$(*,*,*)$	$(0,0,*)$	$(1,0,*)$	$(0,1,*)$	$(1,1,*)$
$(0,0,0)$	.26	.99	0	0	0
$(1,0,0)$	.21	0	.87	0	0
$(0,1,0)$	.21	0	0	.87	0
$(0,0,1)$	0	.01	0	0	0
$(1,1,0)$	.06	0	0	0	.22
$(1,0,1)$	.03	0	.13	0	0
$(0,1,1)$	.03	0	0	.13	0
$(1,1,1)$	.2	0	0	0	.78

Note: Different probabilities (two-digit accuracy) at which the system finds itself, for different clamped units.

*means unclamped. Simulation was run with 1,000 trials. Details in code online in the GitHub folder.

according to equation (10.7), and tally how often units 1 and 2 are active simultaneously. In the small-scale example of table 10.1, I applied the first (direct) approach; this was manageable because one can explicitly calculate all the probabilities $\Pr_w(\mathbf{x})$. However, this explicit approach is feasible only for models with a very small number of units because the number of possible patterns ($\mathbf{x}$) increases exponentially with the number of units.

A second reason why this explicit approach of calculating the probabilities $p_w(i,j)$ (i.e., via probabilities $\Pr_w(\mathbf{x})$) is not feasible is that there can also be hidden units in the Boltzmann model (see chapter 5, and also see figure 8.1b in chapter 8 for illustration). In such a model, we do not have the x_i values available for the hidden units, and therefore we cannot calculate the equilibrium probabilities $\Pr_w(\mathbf{x})$ of specific data patterns $\mathbf{x}$. However, it is obvious from earlier chapters that a neural network can be powerful only if it has hidden units with trainable weights attached to them. For example, the model in figure 10.4a cannot implement the logical XOR rule. We must thus find an efficient approach for sampling the relevant probabilities $p_w(i,j)$ and $p(i,j)$, that appear in the update rule [equation (10.9)]. In the general Boltzmann machine, sampling until equilibrium is reached can be very time-consuming. But a more efficient approach is available by adding restrictions on the connectivity structure, as I discuss in the next section.

Exercises

10.6 Explain as clearly as possible to another person why optimization of equation (10.8) is similar to the cross-entropy maximization in chapter 4.

10.7 Explain how you can obtain probabilities $p_w(i,j)$ based on probabilities $\Pr_w(\mathbf{x})$.

10.8 The text explained how the quantities $p_w(i,j)$ can be calculated. Can you also find out how the quantities $p(i,j)$ are calculated? You can also check the online code in the GitHub folder for inspiration.

Restricted Boltzmann Machines

Consider the model in figure 10.4b. Here, connections are symmetric and only go between layers—from visible to hidden units and vice versa, but not within layers. For this reason, it is called a *restricted Boltzmann machine* (RBM). The interest in this model comes from the fact that for such a restricted model, the learning rule [equation (10.9)] can reasonably be approximated by just a single sample from the relevant $\Pr(\mathbf{x})$ and $\Pr_w(\mathbf{x})$ distributions (Hinton, 2002), or just a few, k, steps; the algorithm is called k-*step contrastive divergence*. This dramatically reduces the time needed for training.

In practice, this goes as follows (for the one-step version of the algorithm). Let's say that the learning rule contains an "environmental" part; that is, $p(i,j)$, data given by the environment. Further, it contains a "model" part; that is, $p_w(i,j)$, the probability that units $\underline{i}$ and j are active together according to the model (parameter w). One clamps an input pattern [sampled from $\Pr(\mathbf{x})$] at the visible units. This pattern is used to determine hidden unit activation as follows:

$$\Pr(x_i^{\text{hid}} = 1) = \frac{1}{1 + \exp\left(-\frac{1}{T}\sum_j w_{i,j}x_j^{\text{vis}}\right)}, \tag{10.10}$$

Unsupervised Learning

where it is now made explicit which units are visible (superscript *vis*) and which are hidden (superscript *hid*). Because the model is restricted (as it is an RBM), hidden units are not connected to other hidden units, and therefore only the visible units are needed to determine if a hidden unit must be on (1) or off (0). The full data pattern $\mathbf{x} = (\mathbf{x}^{vis}, \mathbf{x}^{hid})$ (concatenating input and hidden units) is used to calculate the environmental part of the learning rule [equation (10.9)] (i.e., $p(i,j)$).

What remains to be found are the values of $p_w(i,j)$. We start from the hidden unit activation pattern $\mathbf{x}^{hid}$ from the previous step. This pattern is then projected back to the visible layer, using an activation rule similar to equation (10.10). From the visible layer, activation is projected again back to hidden units using equation (10.10). The last pass is used to calculate the model part of the learning rule [equation (10.9)] (i.e., $p_w(i,j)$). Thus, we have both required quantities $p(i,j)$ and $p_w(i,j)$, and the weights can be updated using equation (10.9). This approach allows the RBM to use hidden units for representing input patterns $\mathbf{x}$. A more detailed summary of this approach can also be found in Testolin et al. (2013).

One application of an RBM are again the logical rules AND, OR, and so on. One can train an RBM with all patterns consistent with the rule and then clamp the visible units to the two input units and evaluate the last, output, visible unit. Table 10.2 shows the result for an RBM that was trained with the AND rule. The GitHub folder for this chapter contains RBM code applied to the MNIST data already discussed in chapters 3 and 4.

Deep RBMs

One important stimulant of the deep learning renaissance of the early twenty-first century was the convolutional model for visual perception, as already discussed in chapter 5 (LeCun et al., 1998). Another stimulant was the development of the RBM, as well as the realization that its hidden units can itself be treated as a "visible" layer. Indeed, once the weights between a visible layer and a hidden layer h_1 are trained (e.g., weights w_1 in figure 10.5a),[6] a pattern from the visible layer can be projected to this

Table 10.2
Restricted Boltzmann machine for implementing the AND rule

Clamped → Pattern x	$(*,*,*)$	$(0,0,*)$	$(1,0,*)$	$(0,1,*)$	$(1,1,*)$
$(0,0,0)$	.07	1	0	0	0
$(1,0,0)$	.19	0	1	0	0
$(0,1,0)$	.34	0	0	1	0
$(0,0,1)$	0	0	0	0	0
$(1,1,0)$	0	0	0	0	0
$(1,0,1)$	0	0	0	0	0
$(0,1,1)$	0	0	0	0	0
$(1,1,1)$	.4	0	0	0	1

Note: Different probabilities (two-digit accuracy) at which the system finds itself, for different clamped units.

*means unclamped. Details in code online.

(a)   **(b)**

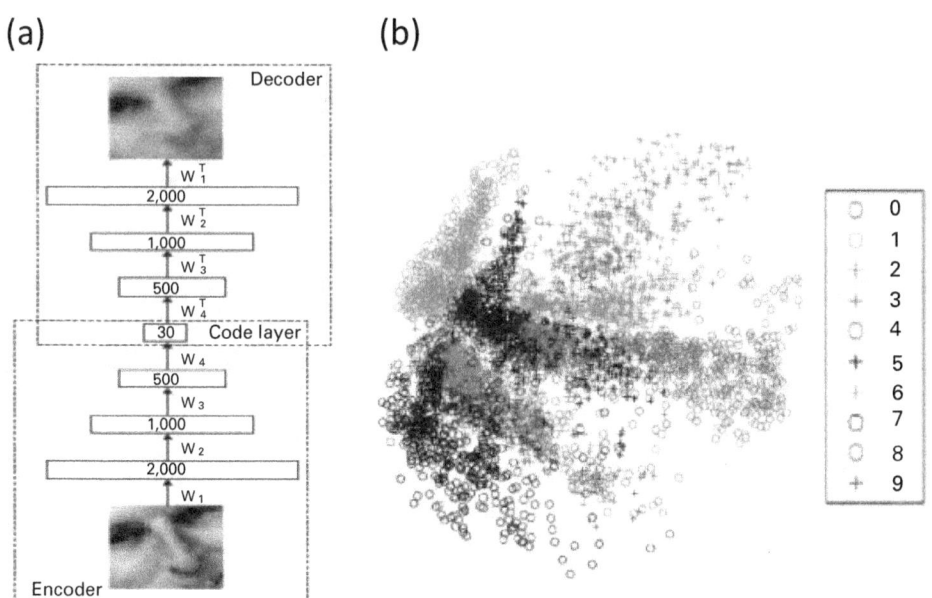

Figure 10.5
Deep RBMs extract structure from data: (a) deep RBM for face recognition; (b) 2D representation of handwritten digits (MNIST) by deep RBM. This figure was reproduced with permission from Hinton and Salakhutdinov (2006).

hidden layer h_1 and the resulting pattern at h_1 can be considered as a visible pattern. Subsequently, the "visible" patterns at h_1 can be projected from layer h_1 to another layer h_2, and the weights between h_1 and h_2 (e.g., weights w_2 in figure 10.5a) can be trained in the same way. Yet another layer can be attached to layer h_2, and so on. In this way, a deep layering of hidden units can be created, with each pair of subsequent layers of units trained using the RBM learning rule.

This was the approach taken by Hinton and Salakhutdinov (2006), which first stacked a number of RBMs in the manner described here. This was followed by unrolling the network, meaning that the deepest layer h_N is mapped again to a novel layer h_{N+1} with weights h_{N-1}, which projects to a novel layer h_{N+2} with weights h_{N-2}, and so on. This process should ideally result in an auto-encoder (see figure 10.5a), where at the final layer (h_{2N-1}), the original input pattern is presented again. Finally, the approach was followed by a few rounds of backpropagation to top it off (i.e., train the weights of the full auto-encoder with a few more steps). The goal of this procedure is obviously not to obtain the activations in the output layer (which are exactly the same as the input layer). The interest resides in the middle layers, which will have optimally compressed the data patterns not linearly, as a principal component analysis would, but in a nonlinear way.

The authors trained their stacked-RBM model on several data sets, including a face database (figure 10.5a) and the MNIST dataset. They observed that after training, the model was able to separate the 10 digits across its hidden units in a highly nonlinear fashion when projected in two dimensions (see figure 10.5b). Thus, it was able to "see

Unsupervised Learning

through the noise" and capture the environmental structure that is available in the 10 digits, and also approximately represent that structure in two dimensions.

Generative Models

Ever since Hermann von Helmholtz, it has been proposed that visual perception is "generative," which means that the function of perception is to construct a model of the world that can explain (generate) the environmental statistics that an organism is confronted with (Gershman & Niv, 2010). For example, at this very moment, millions of photons are hitting my retina. My mind must generate an interpretation of which objects are in the environment (at the moment of writing, laptop screen, table, and so on) that can account for the photon bombardment at the retina.

The Boltzmann machine can implement such a generative model. In an RBM in particular, the function of a higher-level layer is to generate a hypothesis about what goes on in the world. Based on such a hypothesis, this higher-level layer generates an activation pattern on a lower-level layer via the top-down weights (from higher to lower layers). Bottom-up weights (from lower to higher layers) switch on the correct higher-level units (hypotheses). A first hidden layer may implement a number of possible hypotheses simultaneously. This pattern of hypotheses at the first hidden layer can then be again explained by a smaller number of hypotheses at a subsequent level. In this way, retinal (low-level) input can be gradually compressed across deeper layers.

Given that visual perception is to a large degree unsupervised as well, an attractive feature of the generative-model approach is that perception and learning can proceed with minimal requirements from the environment. Based on input patterns from the environment, the model can learn to reconstruct (generate) its own input patterns; for that purpose, it formulates hypotheses (at deeper layers) of what is goes on in the real world. Formulating the right hypothesis allows for knowing the state of the world.

In some sense, the delta rule and backpropagation models are also generative in the sense that they generate a prediction at the output layer, which is subsequently used for learning. However, the aspect of generating (sampling) from a theoretical distribution in order to explain environmental statistics via top-down weights is more explicit in the Boltzmann and other generative models. The idea that the goal of the brain is to generate and explain its own input has been quite influential and has been substantially generalized in recent decades (Friston, 2010).

A related generative model is the Helmholtz machine (Dayan et al., 1995). This model also attempts to make $Pr_w(\mathbf{x})$ as close as possible to $Pr(\mathbf{x})$, and it also has hidden layers at its disposal in order to do so. However, different from the Boltzmann machine, it strictly distinguishes between the top-down (generative) versus bottom-up weights (in other words, w_{ij} is not necessarily equal to w_{ji}). This model is trained with the *wake-sleep algorithm*. As the name suggests, this alternates between a wake phase and a sleep phase: in the wake phase, a data pattern $\mathbf{x}$ is sent bottom-up to the hidden layers, and based on the resulting activation pattern across data and hidden layers, the top-down weights are updated in a gradient-descent step. In contrast, in the sleep phase, activation starts from the hidden layers, is sent top-down to the data layer (as in a dream), and based on the resulting activation pattern in data and hidden layers, bottom-up weights are updated in a gradient-descent step. These steps are alternated until convergence is reached.

Restricted Boltzmann Machines in Cognitive Neuroscience

Whereas Hinton and Salakhutdinov (2006) already applied the RBM architecture to letter perception, later work took this general idea of deep, unsupervised layers that extract successively more structure, as a cognitive neuroscience theory of how humans (and animals) extract structure from the environment. Testolin et al. (2017) applied this idea to letter perception. They trained the first layer of their deep RBM on natural scenes. A subsequent hidden layer was trained only on letters. After training, this layer was found to represent letters (and pseudoletters); moreover, it confused the same letters and fonts as observed in human empirical studies. Interestingly, if the second hidden layer was not based on the natural-scene-based hidden layer 1, its performance was much worse. This suggests that it is useful for the brain to partially recycle the neural structures that it uses for more basic natural-scene processing for the purpose of higher-level symbolic processing (Dehaene & Cohen, 2007).

The same research team studied the development of neural number representations. A recurring theme in numerical cognition is where number representations in the brain originate from; an influential theory is that number representations are innate (Dehaene, 2003; Nieder, 2017). In a layered RBM, Stoianov and Zorzi (2012) presented visual displays to an initially random neural network. They demonstrated that the second hidden layer in the RBM came to spontaneously represent the number of objects in the display in a continuous manner, meaning that a cell is more likely to be active if there are more objects in the environment. Thus, they argued that the number of objects is a useful feature to represent the visual display. The representational scheme observed in the computational model of Stoianov and Zorzi (2012) was consistent with earlier work using human fMRI (Roggeman et al., 2011).

At a more general level, a substantial research literature has studied the idea that visual perception is generative—that is, that perception entails actively predicting the input pattern it receives (see the discussion on generative models earlier in this chapter). The deviations (prediction errors) between prediction and input are typically used to change the weights in such a way as to make better predictions next time. One test of this idea appeared in Egner et al. (2010). In an fMRI experiment, they presented different cues prompting subjects to expect either faces or houses (depending on the cue). At actual stimulus presentation, the authors predicted different trends for house versus face stimuli in face-processing cortical area fusiform face area (FFA), as a function of the cue that was presented. For face stimuli, they predicted less activation (less surprise) in FFA after cues that predicted faces than after cues that predicted houses. In contrast, for house stimuli, they predicted more activation (more surprise) in FFA after cues that predicted faces than after cues that predicted houses. This is indeed the interactive pattern they observed.

Note that this result is very different from what one would expect based on a traditional feedforward processing account of perception. Indeed, in such an account, there would be no activation in face cortical regions at all when a house is presented. Similar prediction-error profiles have been observed in the auditory cortex (Heilbron & Chait, 2018) and in cross-modal (auditory-visual) modalities (den Ouden et al., 2009). The notion of prediction in cognition appears to be ubiquitous in both empirical data and computational models.

To sum up, the goal in unsupervised learning is to represent the original input data from the environment as faithfully as possible. For this purpose, several unsupervised learning models use additional (hidden or output) units with a lower number of dimensions than the original (J-dimensional) input. As a consequence, the network is forced to extract and internally represent the common structure across the input patterns via its hidden or output units and weights.

11 Bayesian Models

When you have eliminated the impossible, whatever remains, however improbable, must be the truth.

—Sherlock Holmes (fictional character by Arthur Conan Doyle; quoted in Kruschke, 2006)

Bayesian models have witnessed a huge popularity in cognitive neuroscience over the last several decades (Behrens et al., 2007; Griffiths et al., 2007). Proponents of these models have sometimes emphasized the difference between them and other cognitive models, such as neural network models (i.e., the supervised learning models discussed in chapters 3–5) (Griffiths et al., 2010). Although there are indeed differences between styles of models, there are also important commonalities. It is my opinion that seeing the commonalities between various modeling approaches is a good way to start appreciating the differences. Here, I will highlight the commonalities between Bayesian models and other approaches to cognitive modeling. Although this chapter is not focused on statistical analysis, I start with a brief overview of Bayesian statistics because that also forms the basis of the Bayesian models.

Bayesian Statistics

The core idea of Bayesian statistics is to place distributions not just on observed variables, as in classical frequentist statistics, but also on parameters that generate those observed variables. Consider a coin that lands heads or tails up with a probability p. We observe variable $X \in \{\text{Heads, Tails}\}$, with the probability of the coin landing heads up being p. In classical (frequentist) statistics, only X has a distribution; in Bayesian statistics, p also has a distribution. For both X and p, Bayesian statisticians interpret the distributions as indicating the amount of belief to be assigned to particular values of the variables (X or p). For example, one can say that the probability that X equals Heads is 0.5; this means that we have a 50% belief that the coin will end heads up. But in Bayesian statistics, we can also make probability statements about parameter p; for example, we could say that the probability that $p = 0.5$ is 0.8. This means that we have an 80% belief that our coin is fair.

Let us denote general distributions and density functions as $f(.)$, although I will use more specific labels for distributions such as Posterior(.) when appropriate. A crucial goal in Bayesian statistics is to derive the distribution of parameters (here, p), conditional on the data (here, X). This distribution is called the *posterior distribution* (with *posterior* referring to "after the data are collected and accounted for"). Bayes's theorem holds that the posterior distribution $f(p|X)$ can be written as follows:

$$f(p|X) = \frac{f(p)f(X|p)}{f(X)}, \qquad (11.1)$$

where $f(p)$ is the *prior* distribution of p (i.e., the distribution of p prior to data collection). The distribution $f(X|p)$ is the probability of the data given parameter p; when considered a function of the parameter p, $f(X|p)$ is just the likelihood function discussed in chapter 6. In this case, where X can take on only two values, its distribution could be a Bernoulli distribution: $f(X = 1 | p) = p$. If one adds a prior distribution for $f(p)$, then equation (11.1) is fully specified; the denominator can be calculated based on the data probability ($f(X|p)$) and prior ($f(p)$), integrated across all possible values of p. A good choice for this prior is a beta distribution (Gelman et al., 1995), which allows us to compute the posterior distribution in equation (11.1).

Examples of the posterior distribution of p [i.e., equation (11.1)] after 5, 50, and 500 coin tosses where the actual probability for Heads equals $p = .8$ are shown in figure 11.1a–c. As a prior, I chose $\alpha = \beta = 2$, meaning that the prior assumes that the coin has no bias; the prior has the highest density on $p = 0.5$ (see the chapter 11 GitHub folder for details; and also see the section entitled "Modeling Streams of Data Using Bayes's Rule," later in this chapter). The alpha and beta as used here represent the parameters of the beta distribution explained later in this chapter: These are not to be confused with the scaling parameters from the earlier chapters. The ordinate $f(p|X)$ shows the density of the posterior belief. To turn this into a probability statement, consider that the belief that the parameter p lies in the interval $(p - \Delta p, p + \Delta p)$ is approximately $2 \times \Delta p \times f(p|X)$.

Intuitively, there should be a strong posterior belief that the true value of p lies in an area with high density $f(p|X)$. For example, after five coin tosses, there is some posterior belief that the value of p should be around .8; and this belief becomes stronger with more data (see figure 11.1a–c). This distribution also allows one to construct $100 \times (1 - \alpha)\%$ belief intervals around specific values. These intervals are indicated with bars on the abscissa in figure 11.1 (with $\alpha = 0.05$, resulting in a 95% belief interval). Here, the larger the sample size, the more the posterior distribution shifts toward higher values of p, away from the prior belief that $p = .5$. Also, the more data there are, the more peaked the distribution, and hence the narrower the belief interval is. Note that we are very likely dealing with an unfair coin (i.e., probability of Heads $\neq .5$) in this case (and indeed, in reality $p = .8$).

Bayes's rule thus provides a recipe for specifying the belief (distribution) about a parameter based on available evidence. A noteworthy feature is that the Bayesian

Bayesian Models 175

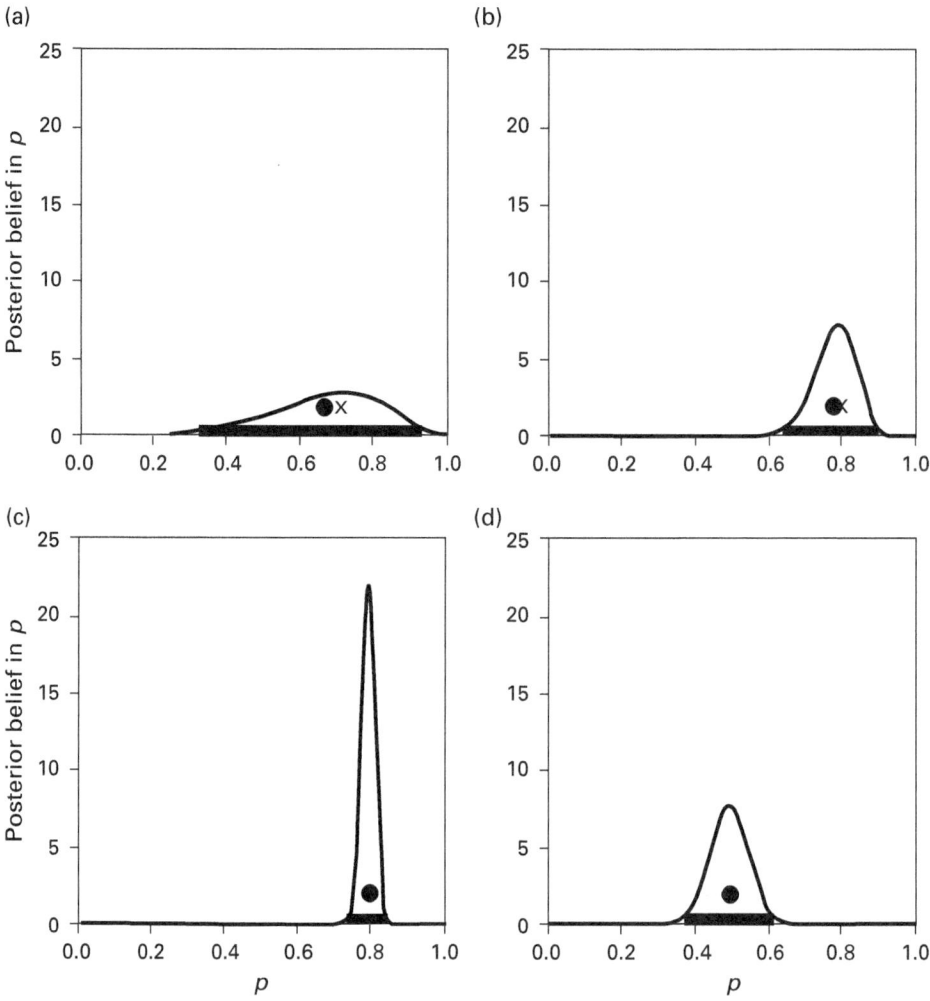

Figure 11.1
Posterior distributions for coin tossing (dot = mean; cross = max of the distribution). The abscissa shows the probability p of the coin landing heads up. Black bars represent 95% belief interval. Panels (a)–(d) show samples 5, 50, 500, and 50, respectively.

approach allows combining prior beliefs with evidence in a natural manner. For figure 11.1a–c, I chose a prior assuming no bias ($\alpha = \beta = 2$); if I instead choose $\alpha = 2$ but $\beta = 20$, there will be a very strong prior belief that p is smaller than .5 (and that the coin is thus biased). Consider figure 11.1d, where I again performed 50 coin tosses (as in figure 11.1b), but now with a strong prior bias $\alpha = 2$, $\beta = 20$, which represents a belief that probability p is low (much smaller than 0.5). The peakedness, and thus certainty, in the distribution are approximately the same as in figure 11.1b; but the distribution has shifted toward the smaller p-values.

Why would one want to incorporate prior beliefs in data collection? One good reason is that earlier data or theory can be formalized as a prior belief. For example, I may already have performed a round of tossing coins before the current round, where I ended up with 0 heads and 18 tails. This earlier data could then be formalized in these new (and biased) prior parameters. Thus, the Bayesian approach allows a natural way to integrate old and novel information. Besides older data, psychological theory (Vanpaemel & Lee, 2012) or an optimal solution (Miynarski et al., 2020) can be incorporated into a prior distribution. However, in many cases, one has no earlier data or prior theory; in such cases, one would choose the prior as uninformative as possible (e.g., $\alpha = \beta = 1$ in the coin tossing example) so that it doesn't influence the conclusions.

Application to the N-Armed Bandit

Just like one can optimize (maximize) a likelihood function as a function of model parameters (as discussed in chapter 6), one can maximize the posterior distribution as a function of those same parameters (i.e., find the mode of the posterior distribution). As a small example, consider again the model for the four-armed bandit problem discussed in chapter 6. Instead of maximizing the likelihood function, I now maximize the posterior distribution of the learning rate and temperature parameters. I have to impose a prior distribution on both parameters and chose a Gaussian distribution with mean = 0 and $sd = 5$. Optimization results (again across 50 replications) can be found in table 11.1. Note that the estimates are pulled toward the prior mean of 0; this process is called *shrinkage* (because it usually involves shrinking the parameter toward zero). Note also that shrinkage is especially severe with a lower number of trials. The more data are observed, the more strongly the likelihood overrules the prior.

Modeling Streams of Data Using Bayes's Rule

In a statistics context, data usually appear in one batch; this was the situation considered up to now. Instead, in a modeling context, it is often more useful to consider the data arriving sequentially, on a trial-to-trial basis. Indeed, we intend our model to mimic cognitive agents, and human agents usually receive information in a sequential manner. Specifically, suppose that at time step t, we have already collected data $X_1, \ldots, X_t$. We can then write the posterior distribution for an arbitrary parameter θ as follows:

$$f(\theta \mid X_1, \ldots X_t) = \frac{f(\theta \mid X_1, \ldots X_{t-1}) f(X_t \mid \theta, X_1, \ldots X_{t-1})}{f(X_t \mid X_1, \ldots X_{t-1})}. \tag{11.2}$$

Table 11.1
Optimizing the posterior distribution of learning rate and temperature parameters

Number of Trials	Learning Rate Mean	Learning Rate Standard Error	Temperature Mean	Temperature Standard Error
100	0.49	0.27	0.66	0.31
1,000	0.62	0.13	0.95	0.11
3,000	0.59	0.09	0.98	0.09

True parameters: learning rate = 0.6; temperature = 1.

Bayesian Models   **177**

We can again recognize the prior and data probability in equation (11.2) at time step t, except that now they are conditioned on $X_1, \ldots, X_{t-1}$. If we ignore the term in the denominator (which is independent of θ), we can rewrite this equation as

$$\text{Posterior}(\theta|X_1, \ldots X_t) \propto \text{Prior}(\theta|X_1, \ldots X_{t-1}) \times \text{Probability}(X_t|\theta, X_1, \ldots X_{t-1}). \quad (11.3)$$

In this way, the posterior belief about the parameter θ can be sequentially updated whenever some information (X_t) arrives. Updating of the posterior can be done based on equation (11.3). To ensure that Posterior(.) is a proper probability distribution, each value $\text{Posterior}(\theta|X_1, \ldots, X_t)$ should be divided by the total mass $\int \text{Posterior}(\theta|X_1, \ldots, X_t)d\theta$ after each update (so that after the division, $\int \text{Posterior}(\theta|X_1, \ldots, X_t)d\theta = 1$).

This updating scheme is often infeasible. For example, if each X_t is a binary variable, then the data probabilities in equation (11.3) condition on 2^{t-1} possibilities for X_1 to X_{t-1}. This is another instance of the curse of dimensionality. However, in many cases, this calculation can be considerably simplified. If the Xs obey the Markov property (as explained in chapter 9), then the data probability becomes $f(X_t|p, X_{t-1})$. In some cases, even X_{t-1} is not needed, so the data probability becomes $f(X_t|p)$. An example of the latter is again the coin-tossing process; the previous tosses $X_1, \ldots X_{t-1}$ are not relevant for the current toss X_t, so we could simply update the posterior distribution as follows after each coin toss at time step t:

$$\text{Posterior}(p|X_1, \ldots X_t) = \text{Prior}(p|X_1, \ldots X_{t-1}) \times p^{X_t}(1-p)^{1-X_t} \quad (11.4)$$

and each value $\text{Posterior}(p|X_1, \ldots, X_t)$ is again divided by $\int \text{Posterior}(p|X_1, \ldots, X_t)dp$ after each time step. This sequential updating based on equation (11.4) is illustrated in figure 11.2. Given that cognitive agents presumably must update information sequentially as well, such sequential updating is an interesting property for models of cognition.

I can now explain the earlier statement that the beta distribution is a good prior distribution in this case. In particular, the beta distribution takes the form

$$f(p|\alpha, \beta) = \frac{p^{\alpha-1}(1-p)^{\beta-1}}{B(\alpha, \beta)}.$$

The denominator is a beta function, which we can ignore; it is data independent and its contribution will be automatically taken care of by the normalization performed after every update. Thus, the contributions from the prior and the data probability can simply be combined in the following way:

$$\text{Posterior}(p|X_1, \ldots X_t) \propto p^{\alpha-1+\Sigma_s X_s}(1-p)^{\beta-1+(t-\Sigma_s X_s)}.$$

One can consider $\alpha - 1$ and $\beta - 1$ to be prior pseudo-observations. Nicely, if $\alpha = \beta = 1$, there are zero pseudo-observations, so then the beta distribution reduces to the (flat) uniform distribution. Earlier data can simply be added to these pseudo-observations thus to construct a novel prior for incoming data. If the prior has this convenient combination property, we say that it is conjugate to the data probability. For discrete (binary) data, the beta prior and Bernoulli probability function are a natural conjugate

178   Chapter 11

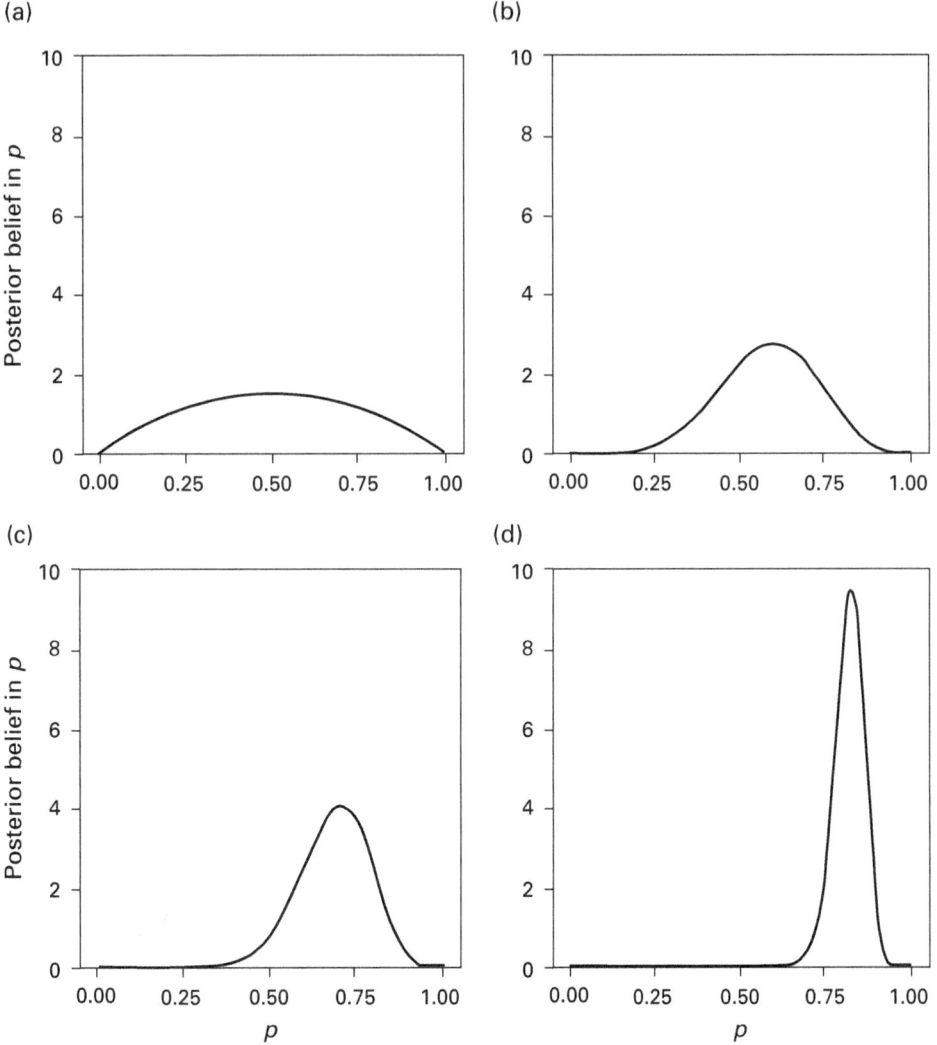

Figure 11.2
Sequential updating of the posterior distribution on parameter p in the coin-tossing example.
Panels (a)–(d) represent posteriors after 2, 10, 20, and 80 trials, respectively.

pair (as just shown); for continuous data, a Gaussian prior and Gaussian density function are a natural conjugate pair for both the mean and the standard deviation (Gelman
et al., 1995).

The Bayesian approach has witnessed a revolution in statistical practice throughout
the sciences, including cognitive neuroscience. One reason for its success is that Bayesian statistics allows for the natural combination of data across subjects and across time
(e.g., different experiments; see the previous discussion). However, given that Bayesian
statistics is extensively considered in other books, I simply refer interested readers to
Dienes (2008), Kruschke (2015), and Lee and Wagenmakers (2013). This chapter instead

Bayesian Models   179

does not concern the use of Bayesian models as models of data, but rather as models of brain and behavior, to which I now turn.

Exercises

11.1 Show how Bayes's rule [equation (11.1)] leads to the update rule [equation (11.2)].

11.2 You want to estimate the probability p of a coin landing heads up. You know absolutely for certain (e.g., from prior studies) that $p > \frac{1}{2}$. How would you formalize this? Would your model or experimental design change? Would you require the same number of observations to achieve a desired level of precision? Implement your model and compare it to the standard (unconstrained) case.

The Rational Approach

From Coin Tossing to Cognitive Science

As mentioned, a key goal in the Bayesian approach is to represent the belief about specific parameters by a statistical distribution. Consider now that a parameter is a very general concept that can apply to any state of the world. So we can model the hypothesis "There is somebody at my doorstep" or "It will rain tomorrow," or indeed any state of the world that one happens to have an interest in. To show this more clearly, I rewrite equation (11.1) in the following way (simply switching terminology):

$$f(\text{Hypothesis} \mid \text{Data}) = \frac{f(\text{Hypothesis}) f(\text{Data} \mid \text{Hypothesis})}{f(\text{Data})}. \qquad (11.5)$$

The Bayesian approach can thus be used to model cognitive hypotheses that a subject may hold about states of the world, how such hypotheses are constructed, and how they are changed based on data that subjects experience. For example, suppose that I model the a priori probability that somebody is at my doorstep to be .001 (people typically don't just stand at my doorstep; they ring the bell or walk by). Further, assume that the probability that I heard the bell ring, given that somebody is there, is .99. We need one more probability to apply Bayes's rule, and that is the probability that I heard the bell ring if nobody is there. Let's say this is .01 (it's not zero—I can hallucinate, or a short circuit may cause a short ring to be emitted). Before listening to the ring, the probability of somebody being there is just the prior probability, .001 (or 0.1%). But now I hear the bell ring! What's the probability that somebody is actually there?

Applying Bayes's rule shows us that the probability has now increased to .09 (or 9%). This may seem to be a bit low; after all, I did hear the bell ring, so somebody must be there (right?). A first reason that this probability turns out lower than expected is that the a priori probability of somebody being there is quite low as well (also only .001). Indeed, if the a priori probability of a visitor were 1% instead of 0.1%, the probability of a visitor given the ringing bell increases to 50%. A second reason is that a ringing bell is not wholly diagnostic of visitor presence (I may hallucinate). This probability was set at 1%; if it is lower (I hallucinate less), then hearing the ringing bell also becomes more diagnostic. For example, if this hallucination probability decreases from 1% to 0.1%, the probability of somebody at my door (given that I heard the bell) also becomes 50%. Readers can increase their intuition about this example, and on Bayesian probabilities more generally, by using the code ch11_bayes.py. So the Bayesian approach prescribes that, given the specified probabilities and data (here, the ringing bell), a subject must update the belief that somebody is at the door from 0.1% to 9%.

Minimizing Loss

Most current Bayesian cognitive models formalize the hypothesis that human perception, action (Körding & Wolpert, 2004, 2006), or cognition (Griffiths & Tenenbaum, 2006) can be understood as application of the Bayesian update equation [equation (11.5)]. But a full Bayesian analysis goes one step further (Anderson, 1991). In particular, one may argue that the purpose of setting up posterior belief distributions about possible states of the world is to formulate an optimal course of action (Huys et al., 2015). To achieve the latter goal, the next logical step is then to define a loss function (i.e., a specification of the loss that each possible action entails, given each possible state of the world). Our goal will then be to minimize the average loss. (A note on terminology: *loss* is also called *cost* in some research domains. Instead of minimizing loss, we can also consider maximizing minus the loss, a quantity which can then be called *value*, as in chapter 9, or *utility*, the term of choice in economics).

For example, suppose that we work at a casino and must decide whether the coin presented to us is a fair one (i.e., probability of Heads =.5), or not. Also suppose that we receive a reward if we guess the actual Heads probability of the coin correctly; otherwise, we receive nothing. This procedure defines our loss function (i.e., the size of the error doesn't matter). In this case, the optimal action to be taken is to choose the parameter value with the highest posterior probability (Mamassian et al., 2001):

$$\hat{p} = \arg\max f(p \mid Data).$$

This optimal parameter is indicated with crosses in figure 11.1.

Instead, suppose that our loss function [denoted $l(a,b)$] is quadratic: we receive money as a function of the squared error, being the difference between our estimate and the real probability p:

$$l(p, \hat{p}) = (p - \hat{p})^2.$$

With this loss function, we receive less money if the squared difference is greater. In this case, the best action to be taken is to choose the estimate $\hat{p}$ to be the mean of the posterior distribution of p given the data:

$$\hat{p} = E(p \mid Data).$$

This optimal parameter corresponding to this loss function is depicted with dots in figure 11.1. The two estimates are usually very similar (and they are exactly the same for symmetric distributions).

Bayesian Models　　　　　　　　　　　　　　　　　　　　　　　　　　**181**

More generally, from an optimality perspective, the point of a Bayesian analysis could be formulated as choosing an action A^* in order to minimize the average loss (Mamassian et al., 2001):

$$A^* = \arg\min_A \int l(\theta, A) p(\theta \,|\, Data) d\theta. \tag{11.6}$$

Bayesian analysis thus naturally complements what is called a *rational* approach to cognition (Anderson, 1991). *Rational* here means just the same as what we called "optimal" in earlier chapters. A rational analysis seeks to identify the best possible course of actions (i.e., A), given the agent's goal (formulated via the loss function) and the agent's environment (Anderson, 1991). The posterior distributions are used so the agent can make predictions about possible states of the world, thus to determine an optimal course of action via equation (11.6).

Although finding the best action A^* via equation (11.6) looks simple in theory, it is typically very hard in practice. Earlier in this chapter, I have already noted the curse of dimensionality problem involved in calculating $p(\theta|\text{Data})$. But there are some additional computational problems here. First, one must integrate over the parameter θ. More important, in calculating the loss $l(\theta, A)$, one must also take into account the possible future consequences of one's actions. This is not an issue in the coin-tossing example because just one action is required. But consider a slightly more complex problem, the N-armed bandit case from earlier chapters. The agent's goal is to make as much money as possible; the agent has N_{trials} trials available to do so (so if each bandit delivers a payoff of either 0 or 1 with an unknown probability, the final payoff will range between 0 and N_{trials}). A rational approach would start from a prior distribution on p [i.e., start from $f(p)$], and then spell out all possible consequences (across all trials). Each combination of choices at each trial can be called a *path* in action space; one possible path is to sample bandit 1 (e.g., leading to a payoff), then sample bandit 2 (e.g., leading to no payoff), sample bandit 2 again (e.g., this time leading to no payoff), and so on. The total number of paths (at trial 1) is $N^{N_{trials}}$; again, an exponential explosion of possible combinations (i.e., the curse of dimensionality). The approach based on equation (11.6) requires calculating a probability for each path. Depending on the specific loss function, an expected outcome (considering both payoffs and costs) is calculated for each path, and the path with the highest expected outcome is chosen. In this case, this could simply consist of choosing the path with the highest expected payoff. After receiving rewards, the distributions $p(\theta|\text{Data})$ are updated, after which the payoffs of $N^{N_{trials}-1}$ paths can be compared on trial 2, and so on until the last trial.

Note that because this approach is rational (i.e., optimal), it also optimally balances exploration and exploitation. For example, the more trials still ahead (N_{trials} is large), the higher will be the expected value of paths containing unexplored options. Thus, the agent should follow those paths containing unexplored options more often than when there are few trials ahead. In the latter case, the agent should instead be more exploitative. For example, if there is just one trial ahead, the agent must be totally exploitative. In an optimal approach, the approximations to trading off exploration versus exploitation from earlier chapters (ε-greedy, softmax, upper confidence bound sampling), are not necessary.

Due to the curse of dimensionality, it is obviously not feasible to actually carry out this algorithm. This is a general issue: as soon as a problem is of some complexity, it is typically not computationally feasible to carry out a rational approach as stipulated here. Thus, a large part of Bayesian and rational analysis consists of making simplifying assumptions to make computationally feasible inferences. In this case, for example, we may only investigate paths up to length $L \ll N_{trials}$, and in the end, choose the path with the highest estimated reward at the end of the path (of length L). By adding such assumptions, one gradually shifts from a rational (but unrealistic) to a psychological (but less unrealistic) model of the agent's behavior. For this reason, the Bayesian modeler typically starts from an optimal perspective, but he then adds extra assumptions to make the model practically manageable. For example, recent theorizing suggests to integrate both rational and resource constraints in modeling (Lieder & Griffiths, 2020). The suboptimal but cognitively realistic model can then be put to the empirical test. We consider some examples of this approach in the next section.

Exercises

11.3 In the bell ringing example, show that the probability of a visitor increases to 9% after the bell rings.

11.4 Show that the identity and square loss functions (as mentioned in the text), lead to the argmax and mean of p as optimal choices (as claimed in the text). Finally, show that with an absolute loss function $l(p, \hat{p}) = |p - \hat{p}|$, the median is the best estimate.

Bayesian Models of Cognition

Perception

The Bayesian approach toward visual perception has a venerable pedigree, dating at least to Hermann von Helmholtz. According to Helmholtz and contemporary Bayesians, people interpret visual scenes by choosing the interpretation that is *most likely*, given the visual input (Mamassian et al., 2001). As a simple example, consider the upper plot of figure 11.3a. One can interpret this scene as a straight line occluded by a box; or alternatively as two lines that just happen to be collinear. However, given that it's unlikely for two lines to end up aligned coincidentally (or that somebody wants to trick you), the visual system would choose the most likely interpretation of a single, occluded line. The effect is a bit weaker in the lower plot of figure 11.3a, where the line is curved (and thus not collinear). In this way, according to Bayesian theorists, the visual system chooses the most likely interpretation of its surrounding visual world.

One natural consequence of Bayesian perception is that subjects combine prior information with data, depending on the amount of uncertainty in each. Consider again equation (11.5); if the information in the likelihood is very precise, the posterior will be strongly pulled toward its maximum. In contrast, if the information in the prior is

Bayesian Models 183

very precise, data will have less of an effect on the subject's belief distribution. This is also illustrated numerically in table 11.1 and graphically in figure 11.1d, where the prior pulls the posterior to the left of the distribution (toward smaller p).

A striking experiment showing that subjects combine sensory with prior information in an uncertainty-weighted manner was reported by Körding and Wolpert (2004). They trained subjects to make finger movements toward a target along a vertical dimension. On each trial, the subject's finger position was revealed to them only once, halfway during the movement. In reality, subjects at that time watched a cursor that was shifted in the horizontal dimension (i.e., perpendicular to movement direction) according to a Gaussian distribution (mean 1 cm). The subjects' goal was to move the cursor to the target. Crucially, the noise at the cursor viewing time was manipulated: subjects could see the cursor not at all (infinite variance), with a lot of noise (high variance), with a small amount of noise (low variance), or perfectly (zero variance). The authors observed that subjects combined prior information about shift size with visual shift information (cursors with different amounts of variance) in a rational manner: The visual information was weighted more strongly when its variance was lower. They concluded that the cognitive system combines visual information weighted by its uncertainty in a near-optimal Bayesian fashion.

As a final example, Meyniel et al. (2015) presented subjects with sequences of stimuli (A or B), where transitions between each two stimuli were governed by the two transition probabilities $Pr(A|A)$ (i.e., the probability of seeing A, given that you've just seen A) and $Pr(B|B)$. At some points in the task, these two probabilities switched both to a novel (i.e., unpredictable) value. Subjects had to estimate these probabilities, and report how confident they were in their judgment. They compared the human data to a

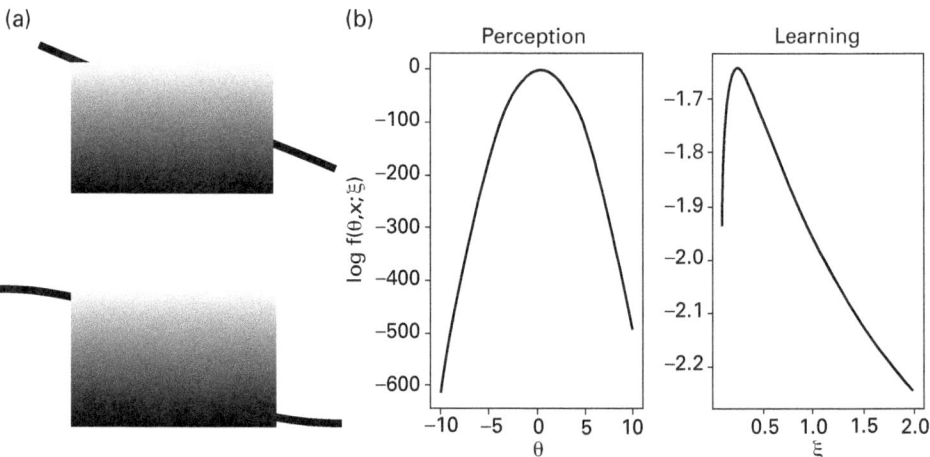

Figure 11.3
Bayesian models in perception and attention. (a) Perception—the upper segments are more often interpreted as resulting from a single line because two line segments lining up like this is unlikely. (b) Joint probability for perception and learning.

so-called Ideal Observer model, which is a model that optimally combines all available data (i.e., the stream of As and Bs) using Bayes's rule. In this example, the Ideal Observer would construct a bivariate posterior distribution for $\Pr(A|A)$ and $\Pr(B|B)$, from which both estimates of these probabilities (mean of the posterior), as well as confidence in these estimates (variance of the posterior) could be obtained. The authors reported that human estimates and confidences corresponded closely to the Ideal Observer model, suggesting that humans can near-optimally estimate such transition probabilities.

Categorization

Categorization also has been considered from a Bayesian perspective. A rational observer model of categorization was proposed by Anderson (1991). Specifically, he proposed how rationally, subjects can judge an object as having value c on feature j, depending on all data seen thus far. For example, feature j could be "has four legs," and the two possible values are in this case yes (the object has four legs) or no (the object does not have four legs). Another feature could be the category an object belongs to (e.g., is it cat or a dog), which clarifies the link to categorization. To make a proper judgment about some feature, one can consider the probability that all objects are partitioned according to some partition Z. A partition divides a set into classes (subsets): for example, with three objects A, B, and C, there are just five possible partitions: {A B C; AB C; AC B; A BC; ABC} (with spaces indicating separation between subsets). One can then compute the probability of a given partition given all data seen thus far, evaluate the probability of the object having value c on feature j under that partition, and then average across all possible partitions. One can thus write the probability that a novel object has value c on feature j as

$$\Pr(X_j = c \mid Data) = \sum_Z \Pr(Z \mid Data)\Pr(X_j = c \mid Z), \tag{11.7}$$

where I introduced a variable X_j that indicates the value of the object on feature j, the *Data* refers to all relevant data, and the sum is taken across all possible partitions. Of course, this is by no means a complete model; both the probabilities on the right side of equation (11.7) need substantial fleshing out before the model becomes plausible or testable. However, there is no need to even begin trying because we already have a curse of dimensionality problem: the number of partitions grows very rapidly as a function of set size.

Anderson then develops an algorithm that does not consider all possible partitions Z, but instead assumes that the agent makes a definite assignment of each object to a subset (or class) k once it has been encountered. For example, one could classify all objects in the subsets "living" or "nonliving." By sequentially assigning each object that is encountered, one specific partition Z is gradually grown (in the "living"/"nonliving" example, the partition would thus have two elements). This assumption dramatically lowers the number of options that needs to be considered. It has accordingly been a very popular assumption in Bayesian models that must assign objects to classes in a cognitively plausible manner (Collins & Frank, 2013). The assumption allows for the

Bayesian Models 185

derivation of a model that specifies the probability of having value c on feature j as follows:

$$\Pr(X_j = c \,|\, Data) = \sum_k \Pr(\text{object} \in k \,|\, \text{features of object}) \Pr(X_j = c \,|\, k). \tag{11.8}$$

Note that the summation on the right side of equation (11.8) is not over the number of partitions as it was in equation (11.7); indeed, as just mentioned, just one partition is tracked. Instead, the summation is over the number of classes of that single partition. This number of classes is gradually incremented as new objects are observed. For each new object, it is decided, based on its feature structure, whether it belongs to one of the already existing classes or instead a new class must be created. Such an assignment additionally requires an equation specifying the a priori probability that the object comes from a novel class. Thus, if the a priori probability of belonging to a novel class is high (the learner is biased toward constructing many classes) or the feature structure of the novel object does not match any of the currently existing classes, a novel class will be created for this object. I refer to the previously cited papers for more information.

The sum across the currently constructed classes k [i.e., in equation (11.8)] consists of two parts. First, there is the probability $\Pr(\text{object} \in k \,|\, \text{features of object})$ that the object belongs in class k, given its feature structure, as discussed in the previous paragraph. Second, there is the probability $\Pr(X_j = c \,|\, k)$ that feature j has a value of c, given that the object is a member of class k. For each of these two probabilities, an incremental update rule is derived based on Bayes's theorem and based on the assumption that features are statistically independent in a class.

Note that a class and a category are defined differently, and thus do not necessarily correspond in this analysis. In fact, class membership is a *latent variable*, and thus unobserved, whereas a category (e.g., is it a cat or a dog) is here just another observable feature (X_j). In fact, there is a strong link with latent variable models from statistics, and in particular latent class models (Vermunt & Magidson, 2004). Just like the neural network models of chapters 4 and 5 conceptualize the agent as performing (non)linear regression, the current Bayesian model also proposes that agents carry out an online version of a standard statistical model: they assign objects to latent classes in order to perform statistical inference on those objects.

Finally, because the category an object belongs to (e.g., cat or dog) is one of the relevant features, one can apply this model to make contact to the categorization literature. Anderson demonstrates that the model corresponds in several respects to empirical data. For example, category structures that humans find hard to learn are also hard to learn for the Bayesian model. Also, objects that show the prototypical structure of a category (think of a prototypical cat for example), will be more easily assigned to that category.

A similar Bayesian model was recently applied to classical conditioning, and in particular to explain context-specificity in conditioning. For example, animals that have learned an association in context A that is unlearned (extinguished) in context B will often reinstate the original association when brought up again in context A. Thus, they express the association in context A, but not in context B. Recent Bayesian models of conditioning propose that animals learn to associate latent causes to events in the environment. The latent causes form a partition Z, as in the categorization model. If contexts A and B are assigned to different latent causes, the animal will activate different associations and thus make different predictions in the two contexts (Courville et al., 2006; Gershman et al., 2010, 2017).

Attention

In daily life, we are constantly bombarded with thousands of stimuli that are far above our capacity to process them all. What should we pay attention to? The nature of attention is one of the classical topics of cognitive psychology. I already briefly considered one implementation of the concept of attention in chapter 8, where it functioned as a credit assignment parameter for learning. A Bayesian model of attention was proposed by Yu and Dayan (2005). They considered the situation where an agent must pay attention to exactly one of N sources. They formalized this situation via a multiple-Posner task. In a standard Posner task, a cue (left- or right-pointing arrow) precedes a target, to which subjects must respond. In the multiple-Posner task, there are instead several cues. The target appears (with a small delay after the cue) either left or right; the cue validity is the probability that the cue correctly predicted the target location. For some number of consecutive trials, one of the sources (cues) predicts the relevant target location, with probability $\eta > \frac{1}{2}$; the other cues predict the correct location at chance level (i.e., with probability $\frac{1}{2}$).

Therefore, the agent must choose which of the N sources to pay attention to. Yu and Dayan (2005) proposed that at trial t, an agent may consider the posterior belief that source j is the relevant source, together with the belief about the current value of η given all data thus far. This posterior probability can be written as

$$\Pr(\text{cue } j \text{ relevant at trial } t, \eta_t | \text{Data}). \tag{11.9}$$

This probability allows for the optimal allocation of attention: simply pay attention to the cue that maximizes equation (11.9). However, calculating this probability is not tractable: it requires that one considers, for all previous $t-1$ trials, the potential relevance of each of the cues at each of those trials, in combination with each possible cue validity level γ. This is the curse of dimensionality again.

Yu and Dayan (2005) then showed that this posterior distribution [equation (11.9)] can be considerably simplified by introducing a few assumptions. They proposed that subjects may just consider a single cue as being the relevant one at each trial t instead of tracking the probability of being relevant for each cue. Subjects could then keep track of the validity of that considered cue (η_t). The quantity $1 - \eta_t$ is thus the unreliability when the identity of the valid cue is known; the authors therefore labeled $1 - \eta_t$ as *expected uncertainty* (EU). If η is low, then EU is high, which means that there is a lot of noise, and therefore subjects also expect that there may be many deviations from their predictions. Simultaneously, subjects may keep track of their confidence that cue j is the relevant one; a parameter that they called λ_t. The parameter $1 - \lambda_t$ was labeled as *unexpected uncertainty* (UU). If subjects have a strong belief in the current cue, then λ is high, UU is low, and they expect that no strong deviations from their predictions will occur.

The authors proposed sequential updating equations for η_t and λ_t. However, under this model, the agent must sometimes make discrete switches. Specifically, at some point, subjects must dispense with cue j and start searching for another cue that is currently relevant. When? The authors demonstrated that subjects (when adhering to their Bayesian model) must assume that the relevant cue has changed whenever

$$UU > \frac{EU}{1/2 + EU}.$$

In other words, *EU* functions as a dynamic bound on UU. If EU is high (i.e., the cue is unreliable), there must be a lot of UU (i.e., strong deviations from their predictions) before subjects will lose faith in the current cue and switch to another cue. In contrast, if EU is low (i.e., the cue is reliable), even a little bit of rule violation may be sufficient for a subject to consider a different cue to be the relevant one.

At the neurobiological level, the authors proposed that EU and UU are signaled by two neuromodulators: EU by acetylcholine (originating from the basal forebrain) and UU by noradrenaline (originating from locus coeruleus). Consistently, the validity effect in the Posner cueing task (where typically just one cue is shown rather than multiple ones, as in this paradigm), is modulated by cholinergic manipulation. Manipulation of noradrenaline, on the other hand, does not modulate the validity effect. In contrast, noradrenaline modulates dimensional shifting when a different dimension becomes relevant in an attentional task (Cain et al., 2011), consistent with a role of noradrenaline in UU. More broadly, an extensive literature suggests a role of noradrenaline for resetting activation in its cortical target structures in unexpected circumstances (Sara & Bouret, 2012).

Note how Bayesian computational models tend to show the same general structure: one starts from a general but cognitively intractable model that is subject to the curse of dimensionality, which is then simplified to a model with tractable and sequential update equations.

Social Cognition

How do we infer the beliefs of other people? Specifically, how does an observer infer what another agent finds rewarding (or costly)? A recent Bayesian model of social cognition postulates that social observers interpret the behavior of other agents by assuming that those (observed) agents optimize value (Jara-Ettinger et al., 2016). Observers would infer the most likely reward (R) and cost (C) structures based on the observations done on a particular subject. The observer sees the agent's actions and then infers what the agents most likely consider to be rewarding (or costly), given the observations. Specifically, social observers would calculate the following posterior probability:

$$\Pr(R, C \,|\, \text{Actions}) \propto \Pr(\text{Actions} \,|\, R, C)\Pr(R, C). \tag{11.10}$$

Once one has the posterior distribution in equation (11.10) available, one can "read the mind" (i.e., the goal function) of the other agent. Specifically, one can find the values of

R and C that maximize the posterior and thus infer what the other agents find reward-ing and costly. The probability $\Pr(\text{Actions}|R, C)$ can be calculated by integrating over all possible policies π; see Jara-Ettinger et al. (2016) for details.

This assumption allowed Jara-Ettinger et al. to account for a host of findings in social cognition. Consider, for example, the Gricean postulates of conversation. Briefly, these postulates stipulate that in communication, agents assume that the other party will share all relevant information, but only that information. For example, suppose that I walk into a restaurant where I'm not consuming anything, ask the bartender, "Do you know where the toilet is?" and the response is "Yes, I do." I'm going to assume then that the bartender does not mean that literally. This would be irrelevant information for the conversation (I already reasonably assume that the bartender has that informa-tion), and thus the bartender does not mean it literally. Instead, the bartender probably means something like "You're not a customer in my restaurant, so I'm not allowing you to use our facilities."

These Gricean postulates follow nicely from the Bayesian model; given that there is a cost for communicating, the other party will only say what is relevant (except perhaps when there is an awkwardness cost of not speaking, as on a date). More experimentally, in a classical experiment, Gergely et al. (2002) demonstrated that preverbal infants will imitate an adult switching on a light using their head, but only when the infants observed that the adults had their hands free. When the adults' hands were bound, children used the lower-cost option of switching on the lights by hand. Their inter-pretation was that children inferred that in the free-hands condition, there must have been a cost for switching on the light with the hands, and therefore adults switched it on using their head. Consequently, the children followed their example in that case. In contrast, in the bound-hands context, it was clear to the children that there was a cost involved in switching on the light with hands, a cost that did not apply to the children themselves (who always had their hands free). Thus, in that condition, they used the (for them, lower-cost) option of switching on the light with their hands. These and many other findings are in line with the model positing that agents optimize a goal function that specifies rewards and costs for those agents, which observers in their turn can estimate using the Bayesian update rule, thus to infer the beliefs of other people.

The Bayesian Brain Hypothesis

Perhaps the most popular Bayesian theory is the Bayesian brain hypothesis put forth by Karl Friston (2009, 2010). To explain the gist of the theory, I summarize here the excellent tutorial of Bogacz (2015). Consider the case where a parameter θ represents a state of the world and leads to an observed variable X; again, visual perception would be the typical example. The parameter θ could represent the real-world angle between two lines; the variable X could represent the angle between the two lines as projected on my retina. I only observe X, but I want to infer the value of θ. In line with the earlier theory, we can write the posterior distribution of θ as

$$\text{Posterior}(\theta|X) \propto f(\theta|X)f(\theta) = f(\theta, X).$$

Bayesian Models   **189**

The latter is called the *joint probability* for the parameter θ and the data X. Therefore, to find the most likely value of θ given the data, we can perform gradient ascent on the log of this joint probability as follows:

$$\Delta\theta = \alpha \frac{\partial}{\partial\theta} \log(f(\theta,X)). \tag{11.11}$$

As in other chapters, repeated application of the gradient ascent rule will tend to maximize the function $f(\theta,X)$. This will give us the best estimate of the state of the world θ, in the sense that it maximizes posterior belief about θ. In a Bayesian framework, this is what *perception* is all about: changing the inferred generative parameters (here, θ) such that the observations (here, X) are as likely as possible.

Now, the distribution $f(\theta,X)$ typically has parameters itself. For example, if θ has a Gaussian prior distribution, parameters of $f(\theta,X)$ would be prior mean and prior variance. Let's call these parameters ξ; so the full form of the joint probability would be $f(\theta,X;\xi)$. Changing these parameters ξ is *learning* in a Bayesian framework. It involves adapting one's cognitive apparatus based on environmental input.

How can one find the optimal parameters ξ? One can again simply perform gradient ascent on the log of the joint probability, as follows:

$$\Delta\xi = \alpha \frac{\partial}{\partial\xi} \log(f(\theta,X;\xi)). \tag{11.12}$$

The negative of $\log(f(\theta,X;\xi))$ is a special case of what is sometimes called *free energy*, for reasons beyond the current discussion (see Buckley et al., 2017). Thus, when equation (11.12) is applied repeatedly, this procedure minimizes free energy in the variable ξ. The left panel of figure 11.3b shows a plot of a joint density for a Gaussian model, as a function of θ; this optimization corresponds to perception. The right panel of the same figure shows the same joint density, but as a function of ξ; this optimization corresponds to learning. For completeness, I note that (11.12) is a special case of negative free energy because I assumed here that both θ and ξ have a distribution with just a single point; in general one can optimize more general distributions (with nonzero probability on more than just a single point) over θ and ξ, in which case the optimization is called variational optimization (Beal, 2003). In that more general case, the free energy can be decomposed in a model fit and a model complexity term.

Exercise

11.5 Show that if prior and data-generating distributions are Gaussian, the update equations [equation (11.11) for data mean, θ and equation (11.12) for prior mean, ξ, respectively] are of a prediction-error format.

Finally, one can also maximize $\log(f(\theta,X;\xi))$ as a function of the variable X. This optimization would correspond to choosing *actions* (a process called *active inference*), such that one samples data (X) that are as likely as possible, given one's model of the world and given one's prior (expectations). Obviously, it is too late to change one's actions in retrospect, so in this case one defines an average free energy across future time steps and chooses actions to minimize that average free energy (Millidge et al.,

2020). Thus, perception, learning, and action can all be formalized as optimization based on the same free energy function.

As I discussed in several earlier chapters, when one algebraically works out the gradient-ascent based update rules for supervised, reinforcement, and unsupervised networks, one ends up with update equations based on *prediction errors* (i.e., differences between actual and predicted outcomes). For several generative models, one also obtained prediction-error based update equations if one works out the current equations [equations (11.11) and (11.12)] (Mathys et al., 2011). Such models are therefore sometimes called *predictive coding models*, although this label is not very discriminative: many reasonable (i.e., goal-optimizing) models require predictions and prediction errors in order to learn.

In summary, the core idea of Bayesian models is that agents sample data from their world in order to construct a posterior belief distribution about parameters that generated those data. As a next step, based on those posteriors, one can choose an action that minimizes a loss function.

12 Interacting Organisms

I will jump into the river to save two brothers or eight cousins
—J. B. S. Haldane

Previous chapters of this book have described various types of interactions that agents may entertain with their environments. An environment could simply provide stimuli without any feedback on an agent's performance (unsupervised learning), it could provide reward feedback (reinforcement learning, or RL), or it could provide detailed labels on what agents are supposed to do next time (supervised learning). The environments that I have considered (with a few exceptions) were indifferent to the agent's actions; they simply said what they had to say, without a goal of their own. However, in social situations, the environment is typically not indifferent. Agents interact with other organisms, and these other organisms have a goal (function) of their own, which may align with the original agent or not.

This chapter considers computational models of social interactions. I discuss three types of social interaction in which computational models have been used. The first can be called "minimally social": agent A makes a single decision toward agent B (or more precisely, multiple statistical independent decisions), and there are no reciprocal actions of B toward A. These assumptions allow a straightforward application of the decision-making tools discussed in chapters 2, 6, and 9 to the field of social decision making. In the second type of interaction, two or more agents combine their knowledge to make a joint decision about a third object (which may itself be social or not). This is briefly considered in the section entitled "Combining Information"; for this purpose, the tools of Bayesian statistics (see chapter 11) can be applied. In the third type of social interaction, iterated and reciprocal interactions (agent A to B, and B to A) are considered. This is considered in the final two sections of the chapter ("Game Theory" and "Cultural Transmission and the Evolution of Languages").

Social Decision Making

In the previous chapters, agents decided between two or more options available to themselves; alternatively, options could also involve a payoff to another agent. I briefly describe two examples of this approach. Hutcherson et al. (2015) presented repeated option choices to their participants, with one fixed option leading to a $50 payoff to both agents and the other option a variable amount (ranging from $10 to $100) to both the agent (denoted $Self) and to the other (denoted $Other). On each trial, the subject decided between the fixed option ($50 for each) and the variable one. The authors next assumed that the relative value $d(t)$ of the two options changes across time steps t (within a trial) as follows:

$$d(t) = d(t-1) + \beta_{\text{Self}}(\$\text{Self} - 50) + \beta_{\text{Other}}(\$\text{Other} - 50) + N(t).$$

When this relative value $d(t)$ reaches an upper bound, the variable option is chosen; when it reaches a lower bound, the fixed option is chosen. Note that this is exactly the decision model from chapter 2, with drift rate $v = \beta_{\text{Self}}(\$\text{Self} - 50) + \beta_{\text{Other}}(\$\text{Other} - 50)$. This is sometimes called a *multiattribute decision model* because the drift rate combines and weights multiple attributes (self and other, in this case). In the experiment, subjects made many such choices, which allowed the authors to fit an accumulation-to-bound model (similar to a diffusion model) to both choice and response time (RT) data (again, see chapter 2). One issue that their model allowed them to address was why generous choices are slower than selfish ones, a finding that earlier research had attributed to generous choices involving a deliberative process (whereas selfish choices would involve a faster, automatic process). Their model predicted the same trend (within the range of the fitted parameters), suggesting that this data pattern cannot be used to support a distinction between an automatic as opposed to a deliberative process (for a similar analysis, see also Krajbich et al., 2015). Moreover, their model predicted (and their data confirmed) the subtler pattern that this effect of slow, generous choices is especially pronounced in more selfish participants.

In another social decision-making paradigm, Crockett et al. (2014) presented their subjects two options for each trial. Each option consisted of a monetary reward, but at a physical cost (namely, an electric shock). Crucially, in some trials, the shock was to be received by the decision maker; in others, the shock was received by another agent instead. Similar to the decision-making models discussed in chapter 6, the authors proposed the following model for acceptance of the shock option:

$$\Pr(\text{Shock}) = \frac{\exp\big(\gamma((1-\kappa)\Delta m - \kappa\Delta s)\big)}{1 + \exp\big(\gamma((1-\kappa)\Delta m - \kappa\Delta s)\big)},$$

where Δm is the difference in monetary reward of the two options and Δs the difference in shock between the two options. The parameter κ indicates how sensitive the participant is for shock versus reward (larger κ, stronger shock-sensitivity). The model had a separate parameter for trials where the subject had to receive the shock himself (κ_{self}), and for trials where another agent was to receive the shock (κ_{other}). Remarkably, the authors observed that on average $\kappa_{\text{other}} > \kappa_{\text{self}}$, suggesting that (at least in this experimental paradigm) harm to others outweighs harm to self.

Combining Information

Another form of social interaction involves combining noisy estimates from different subjects into one combined estimate. For example, different people each may estimate the probability of rain, and an external observer may then decide to bring an umbrella or not, depending on the average estimate. Francis Galton already noted that such averages could yield better estimates than the individual ones on which the average is based (the proverbial wisdom of the crowds). Indeed, if all estimates are generated independently and from the same statistical distributions, the standard error of an average based on N subjects is $N^{1/2}$ times smaller than the standard error based on any single subject. Anecdotally, Galton observed that the average estimate of the weight of an ox (across nearly 800 people) was much more precise than the best individual estimate, and almost identical to the ox's actual weight.

In reality, observations may not be independent, and in such a case, a group decision may become biased. Moreover, even when observations are independent, some estimators may yield better estimates, reflected in different standard deviations for different estimators. As we have seen in chapter 11, from a Bayesian perspective, one can deal with differences between estimators by weighting individual estimates by each estimate's precision (1/variance, or confidence) before combining the estimates. Bahrami et al. (2010) asked whether subjects who make perceptual judgments in pairs would indeed conform to this Bayesian logic, and would weight individual estimates by each estimator's perceptual precision before arriving at a joint decision. Their data suggested that subjects indeed do so; in addition, the authors were able to refute a number of alternative (and less optimal) models, including one that states that the decision of only the subject with the most accurate perceptual judgment would be used in the perceptual decision. They concluded that subjects in that sense have what they called optimally interacting minds.

However, recent research casts doubt on this optimistic conclusion. Indeed, people seem to have an equality bias and do not weight opinions of novices and experts as differentially as would be optimal (Mahmoodi et al., 2015). Also, when making joint perceptual decisions in pairs, the two paired subjects have the tendency to match their average confidence across trials (Bang et al., 2017). Again, this is suboptimal, given that subjects who are better in the perceptual task (i.e., higher precision, higher confidence) should be given greater weight. How such biases are computationally implemented, as well as why humans have them in the first place, currently remain unknown. However, it is useful to have the Bayesian model as a benchmark of what agents could optimally achieve.

Game Theory

I now move to a different line of modeling social interactions called *game theory*, in which iterated interactions between two or more agents, each with its own goal function, are explicitly formalized (von Neumann & Morgenstern, 1944). The word "game" should be interpreted much more broadly than one does in daily life; it concerns any well-defined interaction that two or more agents may have. Depending on the number of players, games are called "1-player," "2-player," or more generally "n-player" games (thus, most of the situations in previous chapters were 1-player games in this sense).

A key goal of game theory is to derive optimal (i.e., reward-maximizing) strategies for different types of games. One very important dimension to distinguish games is whether the summed gain is zero (zero-sum games) or not (nonzero-sum games). If it is zero, then the gain of player A must be fully compensated by an equivalent loss of player B; such games are the most competitive. Nonzero-sum games allow in principle some amount of cooperation between players. Tennis is a zero-sum game: when I win, my opponent necessarily loses (by the same number of games). Economic interaction, on the other hand, is a nonzero-sum game: if I visit a bar, I have a good time, and the barkeep earns a living; both of us profit from the interaction. (However, from a purely financial perspective, this interaction is also a zero-sum game, of course.) A second relevant dimension is whether the game is one-step (just a single play) or iterated. Cooperation is easier to achieve in iterated games.

Let's consider a specific class of 2-player games. Each player has two actions at her disposal, Cooperation (C) and Defection (D), respectively. Depending on the action of each player, a different payoff is given. One of the most celebrated games with this structure is the prisoner's dilemma. Imagine that two criminals are caught and suspected of robbery, but the police only has evidence for trespassing (a less serious crime than robbery). Each criminal is given the option to stay silent (cooperative strategy) or instead betray the other (defecting strategy). When deciding on their choice, they are not allowed to talk to each other. If both betray the other about the robbery, they both go to jail for 2 years. Let's say for simplicity that this outcome has a value (payoff) of –2. If both stay silent, they both go to jail for only 1 year (value of –1). But if only one betrays the other, then the betrayer goes free (value of 0) and the other goes to jail for 3 years (value of –3). The value structure of this game can be formulated as in table 12.1. Different value structures have been used in the literature; but a prisoner's dilemma structure is defined by the value ordering $(D, C) > (C, C) > (D, D) > (C, D)$, where for example (D, C) denotes the outcome if the agent defects (D) and the opponent cooperates (C). Moreover, to avoid that alternating between (D, C) and (C, D) outperforms cooperation

Table 12.1

Value structure for Cooperation or Defection of player A (rows) and player B (columns) in the prisoner's dilemma game

	Player B, C	Player B, D
Player A, C	–1, –1	–3, 0
Player A, D	0, –3	–2, –2

Note: The first and second number in each cell are values for A and B, respectively.

(C, C), one typically also imposes $2(C, C) > (D, C) + (C, D)$ (as is the case in the example here).

In principle, the game's solution is very simple. Each player has a dominant strategy, meaning a strategy that he should do regardless of the other player's choice. This is the Defect action: Both prisoners should betray if they want to maximize their value. As a result, they each will get a value of −2 (2 years in jail). However, there remains the nagging feeling that they could have done better if each cooperated and did not betray the other. They each would have obtained a higher value (−1; just 1 year in jail). And indeed, if the game is played repeatedly for an unknown number of rounds, the best action is not to defect. Intuitively, if the game is played repeatedly, player A cannot hope for cell (D, C) but can avoid cell (C, D). (Given the symmetry of the problem, I'll explain it only from player A's perspective.) Hence, the choice in the repeated-play situation is between (C, C) or (D, D), and here (C, C) is definitely the more valuable one.

But if player A goes for (C, C), how can he avoid ending up in (C, D), the worst position? Generally, player A must attempt to play (C, C), but with sufficient threat to player B to avoid ending up in (C, D). One simple but surprisingly effective strategy to achieve this is called Tit-for-Tat. Here, one starts being cooperative, but as soon as the other player defects, then Tit-for-Tat defects too (Axelrod & Hamilton, 1981). Tit-for-Tat is effective in the sense that it can beat several more sophisticated algorithms, but it is not very robust. For example, suppose that player B has a "trembling hand" and occasionally (in error) defects. If both A and B play Tit-for-Tat, they will defect forever after this error. More robust variants of this algorithm have also been formulated, such as generous-Tit-for-Tat, which cooperates with a fixed probability after a defection by the opponent, but otherwise plays Tit-for-Tat; or Win-Stay-Lose-Shift, which only cooperates after (C, C) and (D, D) outcomes (Nowak & Sigmund, 1993).

Cooperation

The prisoner's dilemma has been so popular because it is at the same time simple and realistic. In several real-world situations, the tendency exists for an individual to defect, which will give that person the largest reward; on the other hand, a cooperative choice will ultimately benefit everyone (including the defector). An old example is that of shepherds providing their sheep access to a pasture shared by several shepherds. Any shepherd can choose to defect and let his own animals overgraze the pasture, leaving nothing for the other shepherds (or future sheep). Named after this example, this problem is sometimes called "the tragedy of the commons" (Hardin, 1968). As a more modern example, consider carbon dioxide (CO_2) emissions. The commons is a clear (breathable) air. When traveling, any individual can choose the cooperative or the defecting choice. A cooperative choice would be travel by bike, in order not to emit too much CO_2, at the cost of spending extra time on travel. The defecting choice would be to travel by car, at a lower time cost for travel. If this game is played repeatedly, the best solution for all players together is definitely that each always cooperates, thus retaining the commons (clean air) that everyone can profit from. But for any individual player, the best choice is to defect (and for everyone else to cooperate). So this dilemma has the same structure as the prisoner's dilemma.

The general problem in iterated prisoner's dilemma games is thus how to enforce cooperative behavior from which everyone ultimately will benefit. This is difficult because reward-maximizing agents are naturally attracted to the (highest reward) defecting choices. Thus, cooperation is hard to achieve, and defection prevails. In well-mixed (i.e., unstructured) populations, cooperators can be exploited by defectors, leading (across iterations) to a population consisting exclusively of defectors. More than just a theoretical question, it is obvious that finding a way to induce cooperation is of great societal interest in these times of looming climate catastrophe.

Computational modeling and mathematical analysis have demonstrated that under special circumstances, cooperation can arise even if each agent strives to maximize his or her self-interest (Nowak, 2006; Rand & Nowak, 2013). Suppose there is a well-mixed population, in which there occurs some (genetic) drift so that strategies can change across time. In this case, Tit-for-Tat players can invade a population of pure-defection strategists. In turn, Tit-for-Tat can be invaded by generous Tit-for-Tat, which itself can be invaded by pure cooperators. Finally, a population of pure cooperators can be invaded by a population of defectors, leading to oscillatory patterns (reminiscent of the oscillations in the Lotka-Volterra model discussed in chapter 1), and in general leading to a constantly varying population (Nowak & Sigmund, 2004). If we are to believe these models, biological agents are rarely purely defecting or cooperative: It depends on the rest of the population.

If the population is not well mixed, more opportunities for cooperation arise (Nowak, 2006). Specifically, so long as agents are sufficiently connected to each other, there is an opportunity for cooperation to emerge between those connected agents. To formulate this argument, we also require the technical assumption that the payoff table is additive, so an agent receives a benefit b if the other person cooperates; but the agent pays a cost c if he or she cooperates (see table 12.2; note that except for a constant shift, table 12.1 also has this additive structure).

The authors distinguish five situations that can lead to cooperation. The first is kin selection. Suppose that when an agent interacts with a relative, any benefit that goes to the relative also goes partially to this agent. The typical example is genetic relatedness; getting b of my genes into the next generation is equivalent to bringing the genes of two of my brothers (each of whom is, approximately, 50% genetically similar to me)

Table 12.2
Additive payoff structure in the prisoner's dilemma game

Player A, Player B	C	D
C	$b-c, b-c$	$-c, b$
D	$b, -c$	$0, 0$

Note: The first and second number in each cell are the payoffs for A and B, respectively.

Interacting Organisms 197

into the next generation because $2 \times \frac{1}{2} \times b = b$. This is why Haldane would jump into the river to save two of his brothers. More generally, if the other player has a relatedness of r, then each benefit b also leads to a benefit $r \times b$ to the original agent. In such cases, Nowak and colleagues showed that cooperation can spread throughout a population (and cannot be invaded by defectors) if $r \times b - c > 0$. Note that the left side of this equation is very similar to value equations from the RL chapters (i.e., chapters 8 and 9), as it is in the format of expected gain − loss > 0.

As an example, kin selection has been used to explain eusociality (strong sociality) in some insect species (e.g., in *Hymenoptera* such as bees or ants). As a result of their mixed genetic status (*Hymenoptera* females are diploid, males are haploid), sisters are 75% genetically related to each other. Thus, the relatedness r is much higher than in more typical, wholly diploid animals (like humans). Therefore, $r \times b - c$ is higher, and as a consequence, cooperation is more beneficial to the individual than it is in diploid species. As a result, these insects would be more cooperative (eusocial) toward their family than diploid species. For example, most female bees and ants do not reproduce themselves, but instead rear the queen's offspring. However, the evolutionary importance of this factor (i.e., being mixed haploid/diploid) is controversial (Johnstone et al., 2011); for example, eusociality also appears in some diploid species.

A second condition that can lead to cooperation is direct reciprocity. Direct reciprocity means that if agent A interacts with agent B, agent A is likely to interact with this same agent B again. Note that this is the case in the repeated prisoner's dilemma game; for example, if the average number of rounds equals 10, the probability of interacting again with the same agent is $p = .91$ (it's less than 1 because the game, and thus the interaction, stops at some time). Suppose that there is a probability p of interacting again with the same agent in this game. Agents play either Tit-for-Tat (a cooperative strategy) or always defect. Under the cooperative strategy, agents are equipped with a memory, and if they encounter somebody who defected against them in an earlier round, they will defect against this agent this time. In all other cases, they cooperate (i.e., play Tit-for-Tat). Nowak and colleagues demonstrated that with this strategy, cooperation pays off[1] under the condition that $p \times b - c > 0$. Again, this makes intuitive sense: if the benefit of cooperation (b) and the probability of playing again (p) are both high relative to the cost of defecting (c), it makes sense to cooperate. As an example of such a Tit-for-Tat strategy, consider vampire bats; they provide food (i.e., blood) to their mates, but specifically only to those that have provided food to them before (Carter & Wilkinson, 2013).

Third, indirect reciprocity can also support cooperation. Suppose that A acts altruistically toward B, and C notices this fact. If the latter increases the probability of a cooperative act of C toward A, then A may benefit from being cooperative toward B. Under a number of technical assumptions, Nowak and colleagues again derived a very similar result as the earlier two. If the probability of knowing the reputation of the other agent is again denoted as p, then cooperation is valuable if $p \times b - c > 0$ (Nowak, 2006).

Fourth, communication structure can also be sufficient for interaction to thrive. This is called *network reciprocity*. Suppose that agents are arranged in a grid, and each agent has exactly k neighbors with whom she can interact. In this case (again under some assumptions), being a cooperator can pay off if $(1/k) \times b - c > 0$. Intuitively, if the number of neighbors is low, islands of agents who cooperate with one another can emerge on the agent interaction grid.

Finally, cooperation can also be defined at the group level. Although controversial for a long time, recent simulations demonstrate that also group-level selection of behavioral strategies (such as cooperation) can operate and drive evolution. Of course, actual selection and reproduction still occurs at the individual level, but if agents in some groups reproduce faster, this can be considered as selection occurring at a higher level than that of the individual. Also for this scenario, a convenient rule can be derived under some simplifying assumptions. In particular, if m denotes the number of groups and N the group size, then group selection for cooperation can occur if $(m/(m + N)) \times b - c > 0$.

Note that all these analyses were derived under the assumption that each agent aims to maximize his own reward. It is not needed to assume that the agents are really concerned with each other. Yet cooperation, with eventual benefit to all agents, can emerge under some circumstances. At the same time, it is clear that cooperation can be precarious, and invasion by defectors remains a constant challenge.

Cultural Transmission and the Evolution of Languages

Let's finally consider an extreme form of information sharing in the animal kingdom: human cultural transmission and change. Referring to figure 1.2 in chapter 1, note that this is the first time that I consider the very slow time scale (i.e., the right side of the x-axis) of changes across generations at the scale of years to centuries. Models of noisy information transmission across generations have also been extensively studied on the even-slower time scale of genetic transmission (Price, 1970); models have also targeted how genes and culture coevolve (Boyd & Richerson, 1985), but these topics are beyond the scope of the current chapter.

A simple but effective metaphor of cultural transmission is the child's game of Telephone, a game where a message is passed from person to person. After a number of iterations, the original message is (often hilariously) deformed relative to the original one. This finding illustrates how noisy transmission across agents can lead to random drift. However, if there is some kind of feedback in the transmission system (e.g., reward or external teaching signals), cultural transmission is typically much more robust and shows interesting regularities.

Properties of cultural transmission can be studied and modeled via the framework of *iterated learning* (Kirby et al., 2014). Here, successive generations of agents pass information (e.g., a language, a cultural habit) to their children or neighbors. To do so, generation n learns a culture from generation $n-1$. Next, generation n subsequently produces a culture that is passed on to generation $n+1$. Individual agents in this information

Interacting Organisms

transmission process can be implemented in various ways, including neural network and Bayesian agents. Another interesting medium in which cultural evolution is studied involves robotic agents, which can learn and interact with either humans or other robots. Discussion of this interesting field is beyond the scope of this book, however (for that, see Cangelosi & Schlesinger, 2015; Steels, 2011).

Next, I will discuss a few of these models and their conclusions on the transmission of language and culture. Note the difference in focus from the earlier chapters: whereas before, I mainly considered change in the individual agents, I now model changes in the products of those agents (i.e., cultural changes, although the agents can also learn in some of the models I will consider). Formally, this means that the goal function changes ownership: whereas before, I considered the goals of the agents, now, the products (e.g., language, cultural habit) generated by these agents have a goal function that can be optimized across several generations of iterated learning. In terms of figure 1.1, interactions at the individual level give rise to transmission and change of information at the social or cultural level above it, a process called *cultural evolution* (Mesoudi, 2011).

One broad conclusion from the iterated learning modeling work in cultural evolution is that, in order for culture to spread, agent density matters. Specifically, simple mathematical models demonstrate that maintenance or increase of cultural complexity requires a sufficiently high density of interacting agents. When the density of simulated agents is too low, their culture dies out more quickly due to copying errors. These models were used to understand several historical events in relation to population density, such as the loss of technical skills in prehistoric Tasmania, which occurred after its disconnection from mainland Australia due to rising sea levels (Henrichs, 2004). Another example is the onset and variability of human culture and cultural transmission (ca. 90,000–20,000 years ago, depending on the region on Earth where the humans lived), and its significant delay relative to modern human anatomy (in Africa at least 160,000 years; (Powell et al., 2009)). For more information on these models and how they were tested and compared to the archaeological record, I refer to Mesoudi (2011).

Learning Language

One particularly important and well-studied form of cultural transmission is language. Iterated learning of neural network models have been used to understand the emergence of several aspects of proto-language, including discrete phonological categories (Oudeyer, 2005), signaling (Cangelosi & Parisi, 1998), and syntax (Cangelosi, 2001). As an example, consider the early model of language transmission and change from Hare and Elman (1995). The authors trained a collection of recurrent neural networks with backpropagation (see chapter 5), on verb inflection of the English language from the Middle Ages (i.e., Old English, ca. 870). Old English was much more irregular than Middle English or than modern English (a more precise quantification of this change appears in Lieberman et al., 2007).

In an iterated learning setup, Hare and Elman (1995) used the output of the neural network in generation n as the target for language learning by the neural network in generation $n+1$. After five simulated generations, they observed a similar shift as empirically observed in the shift from Old English to Middle English; irregular verbs became regularized. Why did this shift occur?

Three aspects drove the regularization. First, words of low *token* frequency (i.e., the word itself is of low frequency) tended to regularize. This can be understood from a neural network perspective; if a word is not presented frequently during training, it has little opportunity to drive the weights in the direction imposed by its target pattern. As a result, the inflection of that word will drift toward the inflection of related words. For the same reason, the past-tense network discussed in chapter 4 regularized irregular words in its second stage. Second, words of low *type* frequency also tended to regularize toward other classes. Similar to the low-token frequency case, if the correct pronunciation of a word is infrequent during training, this pronunciation cannot drive the network weights very strongly and the pronunciation will drift toward that of similar classes of higher type frequency. Finally, the extent to which a verb fits into a verb class (i.e., the verb's consistency) also determines regularization. Verbs that fit well into a class can profit from the weight changes imposed by the other verbs in the class, and thus resist change. Note that no special explanations are needed to explain such change. They follow from iterated learning across agents that are instantiated via standard neural networks, which adhere to the learning principles explained in chapters 4 and 5.

Iterated learning can also be studied with simpler agents. One interesting application concerns the development of grammar across several generations (Kirby, 2001). In particular, this author provided its agents a meaning space (a space of possible messages that one agent may want to pass on to another one); simulated agents had to map this meaning space into a lower-dimensional token space (utterances that can be used to pass on these messages). This bottleneck from a high- to a low-dimensional space turns out to be crucial: with such a bottleneck, agents robustly develop a systematic, compositional language after a few generations of iterated learning. Empirical evidence shows that the same tendency toward compositionality can also happen very rapidly in human language transmission (Kirby et al., 2014). Moreover, when some meanings in the message space were made more frequent, exceptions arose in the language to represent these highly frequent messages more robustly. Consistently, as already mentioned, exceptions also occur in natural language, mostly for high-frequency words.

Another empirical finding considered by iterated learning research is that languages spoken by large populations (e.g., English, Mandarin) show relatively simple grammar but more complex vocabulary. The opposite trend occurs in smaller populations; here, there are relatively fewer words (relative to larger populations), but the grammar is more complex. To understand this phenomenon, Reali et al. (2018) considered a population of simple interacting agents. In their simulations, cultural features were either easy or hard to communicate to other agents. The authors considered easy and hard features to correspond to vocabulary (which is easy to communicate) and grammar (which is harder to communicate), respectively. In the simulations, when the group size was small, information could be transmitted in a relatively reliable manner. As a result, hard-to-communicate features can thrive. In contrast, in larger groups, individual agents only rarely had multiple encounters with the same agent, making communication less reliable. As a result, only easy-to-communicate features such as vocabulary can easily spread throughout the population.

Universal Grammar

A big question in psycholinguistics is whether genetic evolution has endowed humans with an innate and universally applicable grammar (universal grammar). With such a grammar, children must simply set some language-specific parameters and learn the language-specific vocabulary, but the difficult task of learning the syntactic rules is not necessary. Indeed, the universal grammar provides those rules already (Chomsky, 1957). The advantage of this concept of universal grammar, therefore, is that it can explain the incredible speed at which young children can acquire a language.

However, the concept of universal grammar has been extensively criticized. The cognitive scientists Morten Christiansen and Nick Chater have carried out a number of neural network and theoretical studies on the learnability of language (e.g., Christiansen & Chater, 2016a). They simulated the situation where an artificial genome defined the language features of its agents. When the language changed slowly (relative to genetic change), language features eventually became encoded in the genes. This could occur due to a phenomenon known as the Baldwin effect, which is a Darwinian mechanism for passing phenotypic influences onto genes (Hinton & Nowlan, 1987). However, when language changed too quickly, its syntax became a moving target, and could not be genetically encoded by the interacting agents (Chater et al., 2009). Because in reality, language also changes much more slowly than genes, the authors concluded that genetic information is unlikely to encode linguistic features.

If language is not genetically endowed, then how can it be learned so quickly? From the fact that cultural change can occur much more quickly than genetic change, Christiansen and Chater propose that the surprising fit between language and the human brain may be due to changes in language across human generations rather than to changes in the brain (Christiansen & Chater, 2008, 2016b; see also Deacon, 1997). This perspective thus proposes that there may not be an innate, universal grammar, but because language has had the opportunity to evolve toward the biases and assumptions of the human brain, it can be learned very quickly by such brains.

To Conclude

Despite this interesting progress and set of hypotheses that this language modeling work has generated, the languages and interacting agents used in the simulations were obviously highly simplified relative to natural ones. It remains to be shown that the iterated learning perspective accounts for the extreme speed with which human agents can learn and produce complex human languages. This being the conclusion section of this entire book, it is appropriate to generalize this statement. The continuous interaction between modeling and empirical work has led to important insights in cognitive neuroscience, several of which have been discussed in previous chapters. However, several challenges remain.

It is fair to say, for example, that we do not fully comprehend the interactions between levels of cognition (see chapter 1). For example, how does the firing of individual neurons lead (several levels up) to remembering the words of a poem; and how does recitation of this poem charm the person to whom it is addressed? We don't know. One thing that we can be sure of, though, is that in order to fully address this challenge, we will need close interactions and feedback among empirical work, theory development, and computational modeling.

Conventions and Notation

α	Gradient descent scaling parameter
β	Learning rate
γ	Slope
ε	Small value > 0
θ	Threshold
C	An arbitrary constant
E	Energy function
f	Arbitrary function (often, an activation function)
i	Index for output units
I	Number of output units
j	Index for input units or regressors (statistics)
J	Number of variables (e.g., input units)
Log	Natural logarithm
n	Trial number
N	Number of trials (generally, amount of data)
t	Time (typically, in a trial)
unit	The basic building block in a neural network
w_j	Regression coefficient
w_{ij}	Weight from unit j to unit i; units are grouped in layers (so the delta rule works for a two-layer model)
x	Arbitrary variable, activation (input level)
y	Activation at output level
$\leftarrow$	Assignment operator (used in chapter 9)
$\propto$	Is proportional to (used in chapter 11)
$\mathbf{x}^{\mathrm{T}}$	Transpose of vector $\mathbf{x}$ (changes a row to a column or vice versa)

Glossary

Activation function: A transformation carried out on the input of a neural unit.

Argmax: The argument of the maximum of a function. For example, $\mathrm{argmax}([7, 12, -3]) = 2$ because the second element of the list has the maximal value in the list. As another example, $\mathrm{argmax}(-x^2 + 1) = 0$ for $x \in \mathbb{R}$.

Convex set: A set of points in n-dimensional, Euclidean space, such that there is always a linear path between each pair of points in the set. Examples of convex sets are a half-space, a cube, or a sphere.

Delta rule: A supervised learning rule for updating the weights in a two-layer neural network model.

Derivative: Measure of how fast a (univariate) function increases (or decreases) at a fixed point. A derivative of 0 means that the function is constant (neither increasing nor decreasing) at that point.

Discrete: A discrete variable can take on only countably many values (e.g., $X = 1, 2, \ldots$). A continuous variable can be discretized by considering only some of the possible values (e.g., continuous variable X in the range 0 to 1 could be restricted to the points 0, 0.1, 0.2, and so on).

Distributed representations: Neural network representations that distribute the information concerning a specific item, across several units. For example, rather than having one specific unit for the representation of "dog" this concept can be spread across several units. Its opposite is localist representations, where a single unit represents "dog."

Equilibrium distribution: Suppose vectors $\mathbf{x}$ are sampled based on some update rule $\mathbf{x}_n \rightarrow \mathbf{x}_{n+1}$. After repeated sampling, the probability distribution where variables $\mathbf{x}$ are sampled from, will stabilize to some distribution $p(\mathbf{x})$, which is called the equilibrium (or steady-state) distribution.

Gradient: Generalization of the concept of the derivative, for functions of more than one variable. If a function f depends on I variables $x_1, \ldots, x_I$, the elements of the gradient constitute the derivative toward each variable x_i separately. The elements of the gradient are denoted by $\frac{\partial}{\partial x_i} f(x_1, \ldots, x_I)$.

Gradient ascent: An algorithm for ascending (i.e., increasing) a function, making use of the function's gradient.

Gradient descent: An algorithm for descending (i.e., decreasing) a function, making use of the function's gradient.

Logarithm: Function (typically denoted with *log*) with the property that $\log(x \times y) = \log(x) + \log(y)$.

Markov process: An ordered sequence of states where each state depends only on the previous one; in other words, all relevant knowledge for predicting a state, is contained in the previous state.

Matrix: A two-dimensional ordered collection of numbers (e.g., to store network weights).

Neural network: A particular type of computational model. Its goal is to compute some function depending on a collection of (input) variables; each variable is represented by a unit; units form the building block of the neural network. The parameters of the function are typically called weights w_{ij} from unit j to unit i. A synonym (although this term is not used in this book) is connectionist model.

Normality: A vector is said to be normalized if it has a length of 1; algebraically, this means $\mathbf{x}^T\mathbf{x} = \sum_i x_i^2 = 1$.

Orthogonality: Geometrically, two vectors $\mathbf{x}$ and $\mathbf{y}$ are orthogonal if they form a 90-degree angle in Euclidean space; algebraically, this means that $\mathbf{x}^T\mathbf{y} = 0$.

Partial derivative: In a multivariate function (e.g., $f(x,y) = 2x^2 + 3y$), the concept of a (univariate) derivative is generalized to a partial derivative, toward each of the variables (x and y) separately. Its elements are denoted by $\partial f(x,y)/\partial x$ and $\partial f(x,y)/\partial y$ (which, in the example, are $4x$ and 3, respectively). See also *Gradient*.

Policy: A set of rules specifying what action to take in which situation. The goal of reinforcement learning is to find a good (and preferably optimal) policy, in the sense that the policy leads to as much reward as possible.

Policy iteration: A two-step procedure for finding an optimal policy in reinforcement learning. The first step is value estimation (how valuable is each situation or action?); the second step is policy updating (updating one's policy based on the novel value estimates).

Rectification: Cutting off a variable at zero (formally, replace x with $\max(0,x)$).

Value: In reinforcement learning, the value of a state or action is its expected reward from a particular time point on (possibly discounting rewards that occur later in time).

Vector: A one-dimensional, ordered collection of numbers (e.g., the vector $\mathbf{x} = (x_1, x_2, \ldots, x_I)$ may store I network unit values).

Hints and Solutions to Select Exercises

Chapter 1

1.3   Only parameter B influences the optimal point.

1.7   For this function, one can write

$$x_n - B = (1 - 2\alpha A)(x_{n-1} - B).$$

Furthermore, the point x_n will be farther away from the optimum when $|x_n - B| > |x_{n-1} - B|$; because $A > 0$ and $\alpha > 0$, this will be the case when $|1 - 2\alpha A| > 1$, meaning that $2\alpha A > 2$, or $\alpha > 1/A$.

Chapter 2

2.6   Start from

$$E = -in_{\text{cat}} y_{\text{cat}} - in_{\text{dog}} y_{\text{dog}} - wy_{\text{cat}} y_{\text{dog}}.$$

Consider that

$$\frac{dE}{dy_{\text{cat}}} = -\left(in_{\text{cat}} + wy_{\text{dog}}\right),$$

and thus

$$\Delta y_{\text{cat}} = -\alpha \frac{dE}{dy_{\text{cat}}} = \alpha\left(in_{\text{cat}} + wy_{\text{dog}}\right).$$

The update equations follow, with Δy written more explicitly as $y(t) - y(t-1)$, and a noise component $N(t)$ added.

2.9 $E = -w_{21}x_2x_1 - w_{31}x_3x_1 - w_{32}x_3x_2 + \theta_1 x_1 + \theta_2 x_2 + \theta_3 x_3$

2.10

$$E_0 = -\sum_{\substack{j \neq k \\ l \neq k}} \sum_{i < j} w_{i,j} x_i x_j + \sum_{i \neq k} \theta_i x_i$$

$$E_1 = -\sum_{\substack{j \neq k \\ l \neq k}} \sum_{i < j} w_{i,j} x_i x_j + \sum_{i \neq k} \theta_i x_i - \sum_{j \neq k} w_{j,k} x_j + \theta_k,$$

from which the result follows.

208   Hints and Solutions to Select Exercises

Chapter 3

3.1   Consider $E = -\dfrac{1}{I}\sum_{i=1}^{I} t_i y_i$.

Therefore, $\dfrac{\partial E}{\partial w_{ij}} = -\dfrac{1}{I} t_i \dfrac{\partial}{\partial w_{ij}} y_i$. Adding a minus sign (for the steepest descent) and absorbing

the factor $1/I$ into the update rate, one obtains the result.

3.3   The equation can be considered to be a cross-product of a row vector $\mathbf{w}$ and a column vector $\mathbf{x}$, so having multiples of these rows stacked gives the result.

3.4   Consider update rule $\Delta w_{ij} = \beta x_j t_i$. If updates for multiple x_j values are collected in a row vector $\mathbf{x}^T$, the update can be written as $\Delta \mathbf{w}_i = \beta t_i \mathbf{x}^T$. If multiples of these rows are combined, the result follows.

3.5   Under orthonormality of the collection of vectors $\mathbf{x}$,

$$\mathbf{W}\mathbf{x}_k = \sum_{n=1}^{N} \mathbf{t}_n \left(\mathbf{x}_n^T \mathbf{x}_k \right) = \mathbf{t}_k.$$

3.7   One can invent an extra unit $x_{\text{threshold}}$ that always has an activation of 1. This unit has a weight θ_i to unit i. These weights can then be incorporated into the weight matrix $\mathbf{W}$, and the energy equation can (with this extra unit) be written as

$$E = -\sum_{j}\sum_{i<j} w_{i,j} x_i x_j$$

or $E = -\mathbf{x}^T \mathbf{W} \mathbf{x}$.

      In the latter formulation, it would seem that weights are used double (both upper and lower diagonal elements taken). This can be solved by dividing E by 2; zeroing the upper diagonal elements; or placing weights $w_{ij}/2$ instead of w_{ij} in each entry of the matrix.

3.8 $\dfrac{\partial E}{\partial w_{ij}} = -\dfrac{\partial}{\partial w_{ij}} w_{i,j} x_i x_j$

      Multiplying by −1 and adding a learning rate yields the Hebbian learning rule.

3.9   After a single pattern, $\mathbf{W} = \mathbf{x}_1 \mathbf{x}_1^T - \mathbf{I}$. Because $E = -\mathbf{x}^T \mathbf{W} \mathbf{x}$, it follows that $E = -\mathbf{x}^T (\mathbf{x}_1 \mathbf{x}_1^T - I)\mathbf{x}$, which yields the text result (note that $\mathbf{x}$ is J-dimensional and $x_i^2 = 1$ for each i).

3.10 For the Hopfield model, $E = -\mathbf{x}^T \mathbf{W} \mathbf{x}$, as noted in the context of exercise 3.7. For the supervised Hebbian model, $E = -\mathbf{t}^T \mathbf{D}_j \mathbf{y}$, where $\mathbf{D}_j$ is an I-dimensional diagonal matrix with values $1/I$ on the diagonal.

3.11 If a pattern $\mathbf{x}$ is a minimum of this energy function, then the pattern $-\mathbf{x}$ is a minimum also. (Note: This is not true in general, only for this particular energy function.)

Chapter 4

4.1   Under the assumptions stated in the problem, for an activation variable y, one can approximate the difference equation by the following differential equation:

$$\frac{dy}{dt} = \alpha \times (in + wy).$$

      Its solution is

$$y(t) = \frac{in}{-w}\left(1 - e^{\alpha wt}\right).$$

Hints and Solutions to Select Exercises 209

This provides an explicit formulation of y as a function of its input in (and time t). Hence, asymptotically, activation $y(t)$ depends on input (stronger input leads to higher activation) and on leaking (higher leakage, meaning higher $-w$, leads to lower activation).

4.2 Equation (4.2) remains unchanged if one replaces γ by γ/C and simultaneously w by wC for each weight w.

4.3 We start with $\Delta w_{ij} = -\beta \dfrac{\partial E}{\partial w_{ij}}$ [equation (4.5)].

Note that $E = \dfrac{1}{I} \sum_{i=1}^{I} (t_i - y_i)^2$, so we basically have to find

$$\frac{\partial E}{\partial w_{ij}} = \frac{1}{I} \frac{\partial \sum (t_i - y_i)^2}{\partial w_{ij}}, \tag{S.1}$$

where I already moved the constant $1/I$ outside the differentiation operation.

The key is now to see that y_i is a function of the weights w_{ij}, as equation (4.3) shows. We can thus apply the chain rule (check xaktly.com if you don't recall how that works) to derive

$$\frac{\partial E}{\partial w_{ij}} = \frac{\partial E}{\partial y_i} \frac{\partial y_i}{\partial w_{ij}} = -\frac{2}{I}(t_i - y_i)\frac{\partial y_i}{\partial w_{ij}}. \tag{S.2}$$

Note that the summation has disappeared in equation (S.2); this is because deriving E toward w_{ij} in terms that do not contain w_{ij} leads to zero.

The next step is then (again applying the chain rule):

$$\frac{\partial y_i}{\partial w_{ij}} = \frac{df(in_i)}{din_i} \frac{din_i}{dw_{ij}} = \frac{df(in_i)}{din_i} x_j. \tag{S.3}$$

Bringing equations (S.1), (S.2), and (S.3) together into equation (4.5), and absorbing the constant $2/I$ into the constant β, we obtain equation (4.6).

4.4 If each hypothesis is mutually exclusive, we have the following probabilities:

$$p_i = \frac{\exp\left(\sum_j w_{ij} x_j\right)}{\sum_{k=1}^{I} \exp\left(\sum_j w_{kj} x_j\right)}$$

for $i = 1, \ldots, I$; we fix $w_{Ij} = 0$ for all j. (Note that I switched the notation y to p because the latter is more intuitive in this case.)

The log-likelihood for a single data point is in this case written as

$$l = \sum_{i=1}^{I-1} t_i \log(p_i) + t_I \log\left(1 - \sum_{i=1}^{I-1} p_i\right)$$

and thus

$$\frac{\partial l}{\partial w_{kj}} = \sum_{i=1}^{I-1} t_i \frac{1}{p_i} \frac{\partial}{\partial w_{kj}} p_i - t_I \frac{1}{1 - \sum_{i=1}^{I-1} p_i} \frac{\partial}{\partial w_{kj}} \sum_{i=1}^{I-1} p_i \tag{S.4}$$

for $k < I$.

Application of the chain rule shows that

$$\frac{\partial}{\partial w_{kj}} p_k = x_j p_k (1 - p_k) \tag{S.5}$$

and

$$\frac{\partial}{\partial w_{kj}} p_i = -x_j p_i p_k \tag{S.6}$$

when $k \neq i$ and $k, j < I$.

Plugging equations (S.5) and (S.6) into equation (S.4) yields (after some algebraic manipulation) that

$$\frac{\partial l}{\partial w_{kj}} = x_j(t_k - p_k).$$

Hence, the update rule is the same as when the hypotheses are not mutually exclusive.

4.5 $x_1 w_1 + x_2 w_2 > 0$ if and only if $x_2 > -x_1 w_1 / w_2$

4.6 $f(in) = 0.5$ if and only if $x_1 w_1 + x_2 w_2 + \theta = 0$. Thus, the points x_1, x_2 must satisfy a linear equality.

4.7 Stack N input vectors $\mathbf{x}$ as the rows of a matrix $\mathbf{X}$. Then, the output of all data is $\mathbf{XW}^T$ (ignoring the activation function). We now want $\mathbf{XW}^T = \mathbf{t}$. Consider a single output unit; so $\mathbf{W}^T$ is a J-dimensional vector (number of input units), and $\mathbf{t}$ is an N-dimensional vector.

Because input vectors are linearly independent, the square matrix $\mathbf{XX}^T$ will be of full rank (i.e., rank N). Therefore, we can choose as a solution $\mathbf{W}^T = \mathbf{X}^T(\mathbf{XX}^T)^{-1}\mathbf{t}$, and this will reconstruct $\mathbf{t}$ perfectly. This solution is not necessarily unique.

As an aside, if the columns instead of the rows of $\mathbf{X}$ are linearly independent (the more typical case in statistical analysis), then we can choose $\mathbf{W}^T = (\mathbf{X}^T\mathbf{X})^{-1}\mathbf{X}^T\mathbf{t}$. This solution will not necessarily reconstruct $\mathbf{t}$ perfectly, but it is the unique least squares estimator of $\mathbf{W}^T$. If both rows and columns of $\mathbf{X}$ are linearly independent, we can choose $\mathbf{W}^T = \mathbf{X}^{-1}\mathbf{t}$. This solution both perfectly reconstructs $\mathbf{t}$ and is unique.

4.10 Yes

4.11 No

Chapter 5

5.1 If $f_{yx}()$ is linear, then $\mathbf{z} = f_{zy}(\mathbf{W}^{zy} \mathbf{C} \mathbf{W}^{yx}\mathbf{x})$, with $\mathbf{C}$ being the matrix corresponding to the transformation f_{yx}. We can then rewrite this as $\mathbf{z} = f_{zy}(\mathbf{Wx})$, with $\mathbf{W} = \mathbf{W}^{zy} \mathbf{C} \mathbf{W}^{yx}$. So this situation reduces to a single-layer network.

Chapter 6

6.1 The error function can be written in matrix format as $E = (\mathbf{y} - \mathbf{XW})^T(\mathbf{y} - \mathbf{XW})$. Its derivative toward $\mathbf{W}$ (i.e., $\partial E / \partial \mathbf{W}$) equals $-2\mathbf{y}^T\mathbf{X} + 2\mathbf{X}^T\mathbf{XW}$. Setting this derivative to zero and solving toward $\mathbf{W}$ yields the solution.

6.2 Suppose that all variables X and Y are mean-centered. In this case, each element (i,j) of matrix $(\mathbf{X}^T\mathbf{X})$ is just N times the covariance between the ith and jth independent variables. Because the variables in $\mathbf{X}$ are uncorrelated, the off-diagonal elements of this matrix will be zero, and the ith diagonal element of $(\mathbf{X}^T\mathbf{X})^{-1}$ is just $1/(N\sigma^2(X_i))$. Furthermore, the ith element of vector $(\mathbf{X}^T\mathbf{Y})$ is $N\text{Cov}(X_i, Y)$. Given that $\text{Corr}(X_i, Y) = \text{Cov}(X_i, Y) \sigma(X_i) \sigma(Y)$, the result discussed here follows.

Hints and Solutions to Select Exercises 211

6.3 Consider that

$$\frac{\partial \log L}{\partial p} = \frac{n_H}{p} - \frac{n_T}{1-p}.$$

Setting this to zero and solving for p, yields this result.

6.4 Suppose that errors are independent and identically distributed according to a Gaussian distribution with mean zero and standard deviation σ. Then

$$\log L = -N \log(\sqrt{2\pi}\sigma) - \frac{1}{2\sigma^2} \sum_n \left(Y(n) - \sum_j w_j X_j(n) \right)^2.$$

Hence, under these conditions, error minimization and maximum likelihood estimation will yield the same results.

6.5 Suppose that errors are independent and identically distributed according to a Gaussian distribution with mean zero and standard deviation σ. The log-likelihood equals

$$\log L = -N \log(\sqrt{2\pi}\sigma) - \frac{1}{2\sigma^2} \sum_n (Y(n) - \mu)^2.$$

Considering that

$$\frac{\partial \log L}{\partial \mu} = \frac{\sum_i (X_i - \mu)}{\sigma^2},$$

it is clear that the sample mean is the maximum likelihood estimate. This is not generally true; for example, if data are sampled with different noise levels (σ), then data should be weighted by their precision ($1/\sigma^2$) for calculating the maximum likelihood mean.

Chapter 8

8.1 I start writing the expected reward $E(R|W)$, averaged across all possible input patterns $\mathbf{x}$, as

$$E(R|\mathbf{W}) = \sum_{\mathbf{x}} \sum_{\mathbf{y}} \Pr(\mathbf{x}) \Pr(\mathbf{y}|\mathbf{W},\mathbf{x}) E(R|\mathbf{x},\mathbf{y}).$$

The crucial point here is that only $\Pr(\mathbf{y}|\mathbf{W}, \mathbf{x})$ depends on $\mathbf{W}$ (see Williams, 1992). The input pattern $\mathbf{x}$ is sampled by the environment; I can approximate the summation over $\mathbf{x}$ by repeated exposures to the environment (see also exercise 8.2). We will thus approximate the derivative of $E(R|\mathbf{W})$ by random sampling of $E(R|\mathbf{W}, \mathbf{x})$ from $\Pr(\mathbf{x})$. Thus,

$$\frac{\partial E(R|W, \mathbf{x})}{\partial w_{ij}} = \sum_{\mathbf{y}} E(R|\mathbf{x},\mathbf{y}) \frac{\partial}{\partial w_{ij}} \Pr(\mathbf{y}|\mathbf{x},\mathbf{W}).$$

Next, consider that the responses of $\mathbf{y}$ are conditionally independent, given $\mathbf{x}$. Therefore,

$$\frac{\partial}{\partial w_{ij}} \log(\Pr(\mathbf{y}|\mathbf{x},\mathbf{W})) = \frac{\partial}{\partial w_{ij}} \left(\sum_k y_k \log(\Pr(y_k = 1|\mathbf{x},\mathbf{W})) + (1 - y_k) \log(\Pr(y_k = 0|\mathbf{x},\mathbf{W})) \right).$$

Some algebraic manipulations further show that

$$\frac{\partial}{\partial w_{ij}} \log \Pr(y_i = 1|\mathbf{x},\mathbf{W}) = \gamma x_j(1 - p_i)$$

and

$$\frac{\partial}{\partial w_{ij}} \log \Pr(y_i = 0 \,|\, \mathbf{x}, \mathbf{W}) = -\gamma x_j p_i.$$

Note that $y_i \log(\Pr(y_i = 1)) + (1 - y_i)\log(\Pr(y_i = 0))$ can also be written more compactly as $\log(\Pr(y_i))$, so

$$\frac{\partial}{\partial w_{ij}} \log \big(\Pr(y_i \,|\, \mathbf{x}, \mathbf{W})\big) = \gamma x_j (y_i - p_i).$$

Next, using the identity $\partial \log(\Pr(x))/\partial x = (\partial \Pr(x)/\partial x)/\Pr(x)$, it follows that

$$\frac{\partial \Pr(\mathbf{y} \,|\, \mathbf{x}, \mathbf{W})}{\partial w_{ij}} = \gamma \Pr(\mathbf{y} \,|\, \mathbf{x}, \mathbf{W}) x_j (y_i - p_i).$$

Combining all this, we obtain

$$\frac{\partial E(R \,|\, W, \mathbf{x})}{\partial w_{ij}} = \gamma \sum_{\mathbf{y}} E(R \,|\, \mathbf{x}, \mathbf{y}) \Pr(\mathbf{y} \,|\, \mathbf{x}, \mathbf{W}) x_j (y_i - p_i).$$

Thus, the steepest ascent update equation can be obtained.

8.2 In general, we can approximate an integral:

$$\int f(x \,|\, y) g(y)\, dy,$$

if $g(y)$ is a probability (or density) function, by sampling y repeatedly (N times) from $g(y)$ and calculating

$$\frac{1}{N} \sum_{i=1}^{N} f(x \,|\, y_i).$$

(see Tanner, 1996). The result discussed in this chapter is an application of this result.

8.3 $\Pr(y_i = 1) = \dfrac{1}{1 + \sum_{k=1}^{I-1} \exp\left(\gamma \sum_j w_{kj} x_j\right)}$

8.4 This is indeed the case: learning either too slowly or too quickly is suboptimal.

8.5 In your code, you might consider letting γ increase over time [e.g., via the function $\gamma(t) = 1 - \exp(-t)$]. This process is analogous to simulated annealing procedures for learning algorithm optimization (Ackley et al., 1985).

Chapter 9

9.1 An exact answer to this question would require meteorological arguments, so I will formulate this conditionally. If it is the case that the probability of sun after a sunny day is higher in summer than in winter, then the weather does *not* have the Markov property.

9.2 Yes. This question is not even dependent on the previous state, even though that would still be allowed under Markov conditions.

9.3 It seems likely that where you came from should not influence your decision to go left or right as you are going through a maze. If this is the case, then this situation has the Markov property.

Hints and Solutions to Select Exercises   213

9.4   Suppose that we follow a random policy. Then the average amount of temporally discounted reward that we will obtain starting from a cell is exactly this number.

9.5   The value will become more concentrated around the two reward "hot spots."

9.6   The Q-value estimates the average (and temporally discounted) value of choosing action a in state s; but the current update rule only considers the reward in the subsequent state (r), not any later values or rewards. This is appropriate only if temporal discounting is very severe (and in the limit, $\eta = 0$).

Chapter 10

10.1 The parameter β determines the amount to which weight w is pulled toward the novel data point. If $\beta = 0$, it remains where it is; if $\beta = 1$, it ignores everything that it has learned and listens only to the novel data.

10.2 One can write the (hard) competitive learning rule as

$$\Delta w_{ij} = \beta y_i (x_j - w_{ij}),$$

where $y_i = 1$ for the winning unit i, and $y_i = 0$ otherwise. A similar argument holds for the soft competitive (e.g., Kohonen) learning rule.

10.3 Suppose that we want to make $\mathbf{w}$ as close as possible to $\mathbf{x}$. The squared distance between $\mathbf{x}$ and $\mathbf{w}$ equals $(\mathbf{x} - \mathbf{w})^{\mathrm{T}}(\mathbf{x} - \mathbf{w}) = \mathbf{x}^{\mathrm{T}}\mathbf{x} + \mathbf{w}^{\mathrm{T}}\mathbf{w} - 2\mathbf{x}^{\mathrm{T}}\mathbf{w}$. As a function of $\mathbf{w}$, if $\mathbf{w}^{\mathrm{T}}\mathbf{w}$ is constant, we can write this as $(\mathbf{x} - \mathbf{w})^{\mathrm{T}}(\mathbf{x} - \mathbf{w}) = f(\mathbf{w}) = \text{constant} - 2\mathbf{x}^{\mathrm{T}}\mathbf{w}$. Hence, calculating the distance is equivalent to the cross-product between $\mathbf{x}$ and $\mathbf{w}$. Furthermore, note that due to the Cauchy-Schwarz inequality, $\mathbf{x}^{\mathrm{T}}\mathbf{w} \leq (\mathbf{x}^{\mathrm{T}}\mathbf{x} * \mathbf{w}^{\mathrm{T}}\mathbf{w})^{1/2}$. Therefore, we can minimize the distance between $\mathbf{x}$ and $\mathbf{w}$ by choosing $\mathbf{w} = \mathbf{x}$.

It is essential for this argument that $\mathbf{w}^{\mathrm{T}}\mathbf{w}$ is constant rather than $\Sigma\mathbf{w}$. As a simple illustration (from Ten Berge, 2005), suppose that $\mathbf{x} = (2, 1, 1)$; a length ($\mathbf{w}^{\mathrm{T}}\mathbf{w}$) constraint would dictate choosing $\mathbf{w} \propto \mathbf{x}$, but a sum constraint would dictate $\mathbf{w} \propto (1, 0, 0)$.

10.4 If the scale parameter $\tau \to +\infty$, the soft competition becomes gradually more severe, and it eventually ends up as the hard competitive learning rule.

10.5 You can code this and try it. Intuitively, it would mean that you take exactly the opposite steps that the original algorithm prescribes, which is likely to lead to disaster.

10.6 Equation (10.8) is the cross-entropy between $\Pr(\mathbf{x})$ and $\Pr_w(\mathbf{x})$.

10.7 The probability that i and j are jointly active is the sum across all $\Pr_w(\mathbf{x})$ for vectors $\mathbf{x}$ where $x_i = x_j = 1$.

10.8 Tally the observed vectors.

Chapter 11

11.1 Separate $X_1, \ldots, X_{t-1}$ from X_t. Then, apply Bayes's theorem to θ and X_t while conditioning on $X_1, \ldots, X_{t-1}$.

11.2 If you know for certain that $p > \frac{1}{2}$, you could simply truncate all the mass of the prior and posterior distributions at $\frac{1}{2}$. As a result, the mass of the distribution that is lost will be distributed among what remains, and the values with higher posterior will profit more from this redistribution. See the picture here as an illustration. As a result of this redistribution, you can have more posterior belief in your estimates; this is analogous to the increase of statistical power of one-sided hypothesis testing in statistics.

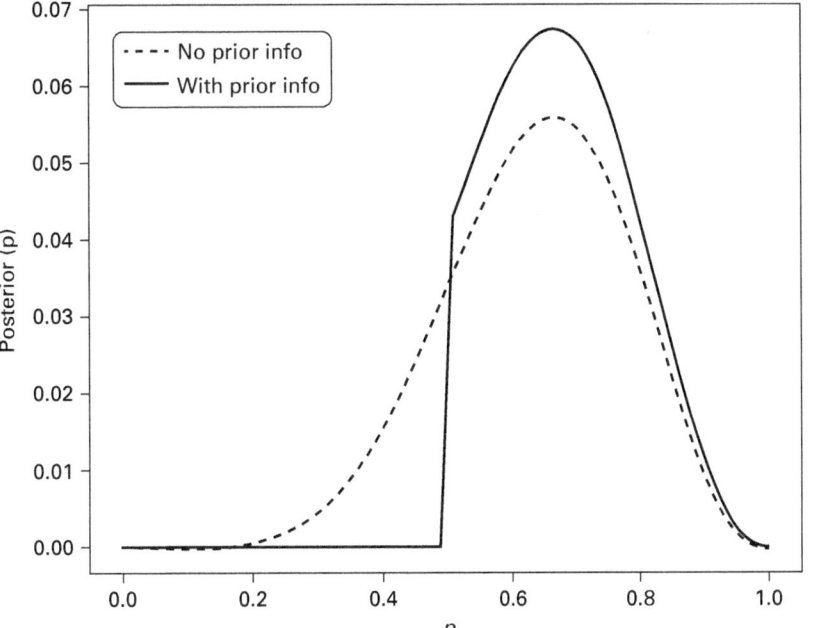

11.3 $\dfrac{0.001 * 0.99}{0.001 * 0.99 + 0.999 * 0.01} = 0.09$

11.4 Consider the loss function:

$$l(g) = \int_{-\infty}^{+\infty} p(x)I(g,x)\,dx,$$

where $I(g,x)$ is 1 if g is the correct x, and 0 otherwise.[1] Then clearly, you want to choose g that maximizes $p(x)$ ($g = \operatorname{argmax}(p(x))$), as mentioned in the chapter.

Next, consider the squared error loss function

$$l(g) = \int_{-\infty}^{+\infty} p(x)(g-x)^2\,dx.$$

Now, simply solve $l'(g) = 0$ toward g, and you will find $g = E(X)$.

Finally, consider the loss function $l(g)$:

$$l(g) = \int_{-\infty}^{+\infty} p(x)|g-x|\,dx,$$

which is equal to

$$l(g) = \int_{-\infty}^{g} p(x)(g-x)\,dx + \int_{g}^{+\infty} p(x)(x-g)\,dx.$$

Now, calculating the derivative of $l(g)$ (using Leibniz' rule of differentiation) and putting that derivative to zero yields

Hints and Solutions to Select Exercises 215

$$\int_{-\infty}^{g} p(x)\,dx = \int_{g}^{+\infty} p(x)\,dx$$

so g must be the median.

11.5 Consider the case where we have data x_i sampled from a normal distribution with mean θ and known standard deviation σ. The mean θ itself follows a normal distribution with mean μ and known standard deviation σ_θ. The log of the posterior distribution then equals (up to a constant):

$$\frac{-(\theta - \mu)^2}{2\sigma_\theta^2} - \frac{\sum_i (\theta - x_i)^2}{2\sigma^2}.$$

We can obtain a step size for μ by deriving toward μ, which yields $(\theta - \mu)/\sigma_\theta^2$, which is in prediction error format.

More interesting is the step size for θ, which is (derive toward θ)

$$\frac{\mu - \theta}{\sigma_\theta^2} + N\frac{\bar{x} - \theta}{\sigma^2}.$$

Thus, θ "listens" to both the prior mean μ and the empirical mean $\bar{x}$, but it weights the two factors depending on amount of data (N) and prior standard deviations (σ_θ, σ).

Notes

Chapter 1

1. You can find the full example here: https://jkkweb.sitehost.iu.edu/jkkteach/Q550/gravmodel .html.

2. This is sometimes called the "computational level," but I use the term "computational" in another sense in this book, so to avoid confusion, I refer to it differently here.

Chapter 2

1. This is to the extent that fact memory is measurable in nonhuman species, of course.

Chapter 3

1. I do not distinguish between short term memory and working memory in this book.

2. The model of Page and Norris is presented as an activation-based model, but can be reformulated as context-based.

Chapter 4

1. In the 1D (figure 4.4a), the line represents the function; in the 2D case (figure 4.4b), the line represents the (x_1, x_2) pairs where the function equals 0.

Chapter 5

1. In fact, there is also a symmetry of solutions in a multilayer network: For example, the weights associated to hidden units 1 and 2 can be completely interchanged, leading to exactly the same error value (see Bishop, 1995, for details). But this is not the issue that regularization deals with.

2. As an aside, yet another class of three-layer models also restricts the feedforward input-to-hidden weights to random values; thus, only the hidden-to-output weights are trained here (Verstraeten et al., 2007). These are called reservoir models; they are in some sense between two- and three-layer models. From two-layer models, they inherit the easy trainability; from three-layer models, they inherit the large space where input features are projected to before mapping to output.

Chapter 6

1. For notational consistency with earlier chapters, I should call this an *I*-armed bandit, but this would violate a stronger constraint of notational consistency with the literature.

2. For a model containing more than one parameter, the second derivative generalizes to the so-called Hessian matrix.

3. A color version of this figure can be found on the MCP website.

Chapter 7

1. In empirical cognitive neuroscience, and typically in electrophysiology research, this is called a *nonparametric approach* because parameters of the statistical distribution do not have to be estimated. I have not used this term, however, to avoid confusion with the term *parameter-free*, discussed earlier in this chapter.

2. Note that the standard practice in fMRI data analysis is to construct different regressors and investigate in which brain areas the activation correlates with which regressors.

Chapter 9

1. See also chapter 6, where a slightly different use of bootstrapping was discussed in the context of statistical model analysis.

2. With Q-learning, this two-step procedure is not necessary because it immediately estimates the optimal value function.

Chapter 10

1. Formally, to do this, one must project each *J*-dimensional data point into the space spanned by the eigenvector corresponding to the largest eigenvalue of the covariance matrix of the input vectors **x**.

2. A color version of figure 10.2 can be found on the MCP website. Note that the online version deviates slightly from the one given here because the latter is optimized for black-and-white printing.

3. A color version of figure 10.3 can be found on the MCP website.

4. Thanks to Naomi Vanderpoorten for suggesting this metaphor.

5. The "distance" meant here is the Kullback-Leibler divergence between the two distributions. It is not strictly speaking a distance function because it is not symmetric in its arguments.

6. A color version of figure 10.5 is available on the MCP website.

Chapter 12

1. Different game-theoretic definitions of the term "payoff" can be defined, for which we refer readers to more specialized literature.

Hints and Solutions to Select Exercises

1. Strictly speaking, this integral is always zero with this loss function; this can be technically solved by changing $I(g, x)$ to a Dirac delta function (beyond the scope of the current text), or by changing the definition of $I(g, x)$ to equal 1 if g is "sufficiently close" to x.

References

Abadi, M., Barham, P., Chen, J., Chen, Z., Davis, A., Dean, J., . . . Zheng, X. (2016). TensorFlow: A system for large-scale machine learning. In *12th USENIX Symposium on Operating Systems Design and Implementation*, 265–283.

Abrahamse, E. L., Braem, S., Notebaert, W., & Verguts, T. (2016). Grounding cognitive control in associative learning. *Psychological Bulletin, 142*(7), 693–728.

Ackley, D. H., Hinton, G. E., & Sejnowski, T. J. (1985). A learning algorithm for Boltzmann machines. *Cognitive Science, 9*, 147–169. https://doi.org/10.1016/S0364-0213(85)80012-4

Anderson, J. A., Silverstein, J. W., Ritz, S. A., & Jones, R. S. (1977). Distinctive features, categorical perception, and probability learning: Some applications of a neural model. *Psychological Review, 84*(5), 413–451.

Anderson, J. R. (1991). The adaptive nature of human categorization. *Psychological Review, 98*(3), 409–429. https://doi.org/10.1037/0033-295X.98.3.409

Anderson, M. C., Bjork, R. A., & Bjork, E. L. (1997). Remembering can cause forgetting: Retrieval dynamics in long-term memory. *Journal of Experimental Psychology: Learning Memory & Cognition, 20*(5), 1063–1087.

Annis, J., Gauthier, I., Palmeri, T. J., Annis, J., Gauthier, I., & Palmeri, T. J. (2020). Combining convolutional neural networks and cognitive models to predict novel object recognition in humans. *Journal of Experimental Psychology: Learning, Memory, and Cognition Learning Memory and Cognition.* https://doi.org/10.1037/xlm0000968

Ashby, F. G., Ennis, J. M., & Spiering, B. J. (2007). A neurobiological theory of automaticity in perceptual categorization. *Psychological Review, 114*(3), 632–656. https://doi.org/10.1037/0033 -295X.114.3.632

Averbeck, B. B., Chafee, M. V, Crowe, D. A., & Georgopoulos, A. P. (2002). Parallel processing of serial movements in prefrontal cortex. *Proceedings of the National Academy of Sciences, 99*, 13172–13177.

Avillac, M., Denève, S., Olivier, E., Pouget, A., & Duhamel, J.-R. (2005). Reference frames for representing visual and tactile locations in parietal cortex. *Nature Neuroscience, 8*(7), 941–949. https://doi.org/10.1038/nn1480

Axelrod, R., & Hamilton, W. D. (1981). The evolution of cooperation. *Science, 211*(4489), 1390–1396.

Baddeley, A., & Hitch, G. J. (1974). Working memory. In G. D. Bower (Ed.), *The psychology of learning and motivation* (pp. 47–89). Academic Press.

Bahrami, B., Olsen, K., Latham, P. E., Roepstorff, A., Rees, G., & Frith, C. D. (2010). Optimally interacting minds. *Science, 329*(5995), 1081–1086.

Bakker, A., Kirwan, C. B., Miller, M., & Stark, C. E. L. (2008). Pattern separation in the human hippocampal CA3 and dentate gyrus. *Science, 319*(5870), 1640–1642. https://doi.org/10.1126/science.1152882

Ballard, I. C., & McClure, S. M. (2019). Joint modeling of reaction times and choice improves parameter identifiability in reinforcement learning models. *Journal of Neuroscience Methods, 317*(April), 37–44. https://doi.org/10.1016/j.jneumeth.2019.01.006

Bang, D., Aitchison, L., Moran, R., Castanon, S. H., Rafiee, B., Mahmoodi, A., . . . Summerfield, C. (2017). Confidence matching in group decision-making. *Nature Human Behaviour, 0117*(May), 1–7. https://doi.org/10.1038/s41562-017-0117

Bassett, D. S., Bullmore, E., Verchinski, B. A., Mattay, V. S., Weinberger, D. R., & Meyer-Lindenberg, A. (2008). Hierarchical organization of human cortical networks in health and schizophrenia. *Journal of Neuroscience, 28*(37), 9239–9248. https://doi.org/10.1523/JNEUROSCI.1929-08.2008

Bays, P. M., & Husain, M. (2008). Dynamic shifts of limited working memory resources in human vision. *Science, 321*, 851–854. https://doi.org/10.1126/science.1158023

Beal, M. J. (2003). Variational algorithms for approximate Bayesian inference. University College London.

Behrens, T. E. J., Woolrich, M. W., Walton, M. E., & Rushworth, M. F. S. (2007). Learning the value of information in an uncertain world. *Nature Neuroscience, 10*(9), 1214–1221. https://doi.org/10.1038/nn1954

Bishop, C. C. (1995). *Neural networks for pattern recognition*. Oxford University Press.

Bishop, Y. M., Fienberg, S. E., & Holland, P. W. (2007). *Discrete multivariate analysis: Theory and applications*. Springer.

Bogacz, R. (2015). A tutorial on the free-energy framework for modelling perception and learning. *Journal of Mathematical Psychology, 76,* Part B, 198–211. https://doi.org/10.1016/j.jmp.2015.11.003

Bogacz, R., Wagenmakers, E.-J., Forstmann, B. U., & Nieuwenhuis, S. (2010). The neural basis of the speed-accuracy tradeoff. *Trends in Neurosciences, 33*(1), 10–16. https://doi.org/10.1016/j.tins.2009.09.002

Botvinick, M. M., Braver, T. S., Barch, D. M., Carter, C. S., & Cohen, J. D. (2001). Conflict monitoring and cognitive control. *Psychological Review, 108*(3), 624–652.

Botvinick, M. M., & Plaut, D. C. (2004). Doing without schema hierarchies: A recurrent connectionist approach to normal and impaired routine sequential action. *Psychological Review, 111*(2), 395–429. https://doi.org/10.1037/0033-295X.111.2.395

Bourne, J. N., & Harris, K. M. (2011). Coordination of size and number of excitatory and inhibitory synapses results in a balanced structural plasticity along mature hippocampal CA1 dendrites during LTP. *Hippocampus, 21*(4). https://doi.org/10.1002/hipo.20768

References 221

Boyd, R., & Richerson, P. J. (1985). *Culture and the evolutionary process*. University of Chicago Press.

Bromberg-Martin, E. S., & Hikosaka, O. (2009). Article midbrain dopamine neurons signal preference for advance information about upcoming rewards. *Neuron, 63*(1), 119–126. https://doi.org/10.1016/j.neuron.2009.06.009

Brown, G. D. A., Preece, T., & Hulme, C. (2000). Oscillator-based memory for serial order. *Psychological Review, 107*(1), 127–181. https://doi.org/10.1037/0033-295X.107.1.127

Brown, J. W., & Braver, T. S. (2005). Learned predictions of error likelihood in the anterior cingulate cortex. *Science, 307*(5712), 1118–1121. https://doi.org/10.1126/science.1105783

Brown, S. D., & Heathcote, A. (2008). The simplest complete model of choice response time: Linear ballistic accumulation. *Cognitive Psychology, 57*(3), 153–178. https://doi.org/10.1016/j.cogpsych.2007.12.002

Buckley, C. L., Sub, C., McGregor, S., & Seth, A. K. (2017). The free energy principle for action and perception: A mathematical review. *Journal of Mathematical Psychology, 81*, 55–79. https://doi.org/10.1016/j.jmp.2017.09.004

Burgess, N., & Hitch, G. J. (1999). Memory for serial order: A network model of the phonological loop and its timing. *Psychological Review, 106*(3), 551–581.

Burgess, N., & Hitch, G. J. (2005). Computational models of working memory: Putting long-term memory into context. *Trends in Cognitive Sciences, 9*(11), 535–541. https://doi.org/10.1016/j.tics.2005.09.011

Busemeyer, J. R., Myung, I. J., & McDaniel, M. A. (1993a). Cue competition effects: Empirical tests of adaptive network learning models. *Psychological Science, 4*(3), 190–196.

Busemeyer, J. R., Myung, I. J., & McDaniel, M. A. (1993b). Cue competition effects: Theoretical implications for adaptive network learning models. *Psychological Science, 4*(3), 196–203.

Butz, M. V., & Kutter, E. F. (2017). *How the mind comes into being: Introducing cognitive science from a functional and computational perspective*. Oxford University Press.

Cain, R. E., Wasserman, M. C., Waterhouse, B. D., & McGaughy, J. A. (2011). Atomoxetine facilitates attentional set shifting in adolescent rats. *Developmental Cognitive Neuroscience, 1*(4), 552–559. https://doi.org/10.1016/j.dcn.2011.04.003

Calderon, C. B., Dewulf, M., Gevers, W., & Verguts, T. (2017). Continuous track paths reveal additive evidence integration in multistep decision making. *Proceedings of the National Academy of Sciences, 114*, 10618–10823.

Calderon, C. B., Gevers, W., & Verguts, T. (2018). The unfolding action model of initiation times, movement times, and movement paths. *Psychological Review, 125*(5), 785–805.

Cangelosi, A. (2001). Evolution of communication and language using signals, symbols, and words. *IEEE Transactions on Evolutionary Computation, 5*(2), 93–101.

Cangelosi, A., & Parisi, D. (1998). The emergence of a "language" in an evolving population of neural networks. *Connection Science, 10*(2), 83–98.

Cangelosi, A., & Schlesinger, M. (2015). *Developmental robotics: From babies to robots*. MIT Press.

Caporale, N., & Dan, Y. (2008). Spike timing-dependent plasticity: A Hebbian learning rule. *Annual Review of Neuroscience, 31*, 25–46. https://doi.org/10.1146/annurev.neuro.31.060407.125639

Carter, G. G., & Wilkinson, G. S. (2013). Food sharing in vampire bats: Reciprocal help predicts donations more than relatedness or harassment. *Proceedings of the Royal Society B*, February 22. https://doi.org/10.1098/rspb.2012.2573

Chater, N., Reali, F., & Christiansen, M. H. (2009). Restrictions on biological adaptation in language evolution. *Proceedings of the National Academy of Sciences, 106*(4), 1015–1020.

Chatham, C. H., Frank, M. J., & Badre, D. (2014). Corticostriatal output gating during selection from working memory. *Neuron, 81*(4), 930–942. https://doi.org/10.1016/j.neuron.2014.01.002

Cheadle, S., Wyart, V., Tsetsos, K., Myers, N., de Gardelle, V., Herce Castañón, S., & Summerfield, C. (2014). Adaptive gain control during human perceptual choice. *Neuron, 81*(6), 1429–1441. https://doi.org/10.1016/j.neuron.2014.01.020

Chen, Q., & Mirman, D. (2012). Competition and cooperation among similar representations: Toward a unified account of facilitative and inhibitory effects of lexical neighbors. *Psychological Review, 119*(2), 417–430. https://doi.org/10.1037/a0027175

Chomsky, N. (1957). *Syntactic structures*. Mouton & Co.

Chomsky, N. (1959). Review of B. F. Skinner's *Verbal Behavior. Language, 35*, 26–58.

Christiansen, M. H., & Chater, N. (2008). Language as shaped by the brain. *The Behavioral and Brain Sciences, 31*(5), 489–508; discussion 509–558. https://doi.org/10.1017/S0140525X08004998

Christiansen, M. H., & Chater, N. (2016a). *Creating language*. MIT Press.

Christiansen, M. H., & Chater, N. (2016b). The Now-or-Never bottleneck: A fundamental constraint on language. *Behavioral and Brain Sciences, 39*.

Cisek, P., Puskas, G. A., & El-Murr, S. (2009). Decisions in changing conditions: The urgency-gating model. *Journal of Neuroscience, 29*(37), 11560–11571. https://doi.org/10.1523/JNEUROSCI.1844-09.2009

Cogliati-Dezza, I., Cleeremans, A., & Alexander, W. (2019). Should we control? The interplay between cognitive control and information integration in the resolution of the exploration-exploitation dilemma. *Journal of Experimental Psychology: General, 148*(6), 977–993.

Collins, A. G. E., & Frank, M. J. (2013). Cognitive control over learning: Creating, clustering, and generalizing task-set structure. *Psychological Review, 120*(1), 190–229. https://doi.org/10.1037/a0030852

Collins, A. G. E., & Frank, M. J. (2018). Within- and across-trial dynamics of human EEG reveal cooperative interplay between reinforcement learning and working memory. *Proceedings of the National Academy of Sciences*, 1–6. https://doi.org/10.1073/pnas.1720963115

Coltheart, M., Rastle, K., Perry, C., Langdon, R., & Ziegler, J. (2001). DRC model: A dual route cascaded model of visual word recognition and reading aloud. *Psychological Review, 108*(1), 204–256.

Cooper, L. N., & Bear, M. F. (2012). The BCM theory of synapse modification at 30: Interaction of theory with experiment. *Nature Reviews Neuroscience, 13*(11), 798–810. https://doi.org/10.1038/nrn3353

Cooper, R. P., Ruh, N., & Mareschal, D. (2014). The goal circuit model: A hierarchical multi-route model of the acquisition and control of routine sequential action in humans. *Cognitive Science, 38*, 244–274. https://doi.org/10.1111/cogs.12067

References 223

Cooper, R. P., & Shallice, T. (2006). Hierarchical schemas and goals in the control of sequential behavior. *Psychological Review, 113*(4), 887–916. https://doi.org/10.1037/0033-295X.113.4.887

Courville, A. C., Daw, N. D., & Touretzky, D. S. (2006). Bayesian theories of conditioning in a changing world. *Trends in Cognitive Sciences, 10*(7), 294–300. https://doi.org/10.1016/j.tics.2006.05.004

Cowan, N. (2001). The magical number 4 in short-term memory: A reconsideration of mental storage capacity. *Behavioral and Brain Sciences, 24*(1), 87–114. https://doi.org/10.1017/S0140525X01003922

Crockett, M. J., Kurth-Nelson, Z., Siegel, J. Z., Dayan, P., Dolan, R. J., Crockett, M. J., . . . Dolan, R. J. (2014). Harm to others outweighs harm to self in moral decision making. *Proceedings of the National Academy of Sciences, 111*(48), 17320–17325.

Dale, A. M., & Buckner, R. L. (1997). Selective averaging of rapidly presented individual trials using fMRI. *Human Brain Mapping, 5*(5), 329–340. https://doi.org/10.1002/(SICI)1097-0193(1997)5:5<329::AID-HBM1>3.0.CO;2-5

Dandurand, F., Grainger, J., & Dufau, S. (2010). Learning location-invariant orthographic representations for printed words. *Connection Science, 22*(1), 25–42. https://doi.org/10.1080/09540090903085768

D'Ardenne, K., McClure, S. M., Nystrom, L. E., & Cohen, J. D. (2008). BOLD responses reflecting dopaminergic signals in the human ventral tegmental area. *Science, 319*(5867), 1264–1267. https://doi.org/10.1126/science.1150605

Davelaar, E. J., Goshen-Gottstein, Y., Ashkenazi, A., Haarmann, H. J., & Usher, M. (2005). The demise of short-term memory revisited: Empirical and computational investigations of recency effects. *Psychological Review, 112*(1), 3–42. https://doi.org/10.1037/0033-295X.112.1.3

Davis, C. J. (2010). The spatial coding model of visual word identification. *Psychological Review, 117*(3), 713–758. https://doi.org/10.1037/a0019738

Daw, N. D. (2009). Trial-by-trial data analysis using computational models. In M. R. Delgado, E. A. Phelps, & T. W. Robbins (Eds.), *Decision making, affect, and learning. Attention & performance XXIII* (pp. 1–26). Oxford University Press Scholarship Online.

Daw, N. D., Gershman, S. J., Seymour, B., Dayan, P., & Dolan, R. J. (2011). Model-based influences on humans' choices and striatal prediction errors. *Neuron, 69*(6), 1204–1215. https://doi.org/10.1016/j.neuron.2011.02.027

Daw, N. D., Niv, Y., & Dayan, P. (2005). Uncertainty-based competition between prefrontal and dorsolateral striatal systems for behavioral control. *Nature Neuroscience, 8*(12), 1704–1711. https://doi.org/10.1038/nn1560

Dawkins, R. (1976). *The selfish gene*. Oxford University Press.

Dayan, P., Hinton, G. E., Neal, R. M., & Zemel, R. S. (1995). The Helmholtz machine. *Neural Computation, 7*(5), 889–904. http://www.ncbi.nlm.nih.gov/pubmed/7584891

Deacon, T. (1997). *The symbolic species: The co-evolution of language and the brain*. W. W. Norton.

Deci, E. L., & Ryan, R. M. (2008). Self-determination theory: A macrotheory of human motivation, development, and health. *Canadian Psychology, 49*(3), 182–185. https://doi.org/10.1037/a0012801

Dehaene, S. (1997). *The number sense: How the mind creates mathematics*. Oxford University Press.

Dehaene, S. (2003). The neural basis of the Weber-Fechner law: A logarithmic mental number line. *Trends in Cognitive Sciences, 7*(4), 145–147. http://www.ncbi.nlm.nih.gov/pubmed/12691758

Dehaene, S., & Cohen, L. (2007). Cultural recycling of cortical maps. *Neuron, 56*(2), 384–398. https://doi.org/10.1016/j.neuron.2007.10.004

Dehaene, S., & Mehler, J. (1992). Cross-linguistic regularities in the frequency of number words. *Cognition, 43(1)*, 1–29. http://www.sciencedirect.com/science/article/pii/001002779290030L

De Loof, E., Ergo, K., Naert, L., Janssens, C., Talsma, D., Van Opstal, F., & Verguts, T. (2018). Signed reward prediction errors drive declarative learning. *PloS One, 13(1)*, e0189212.

Deneve, S., & Pouget, A. (2003). Basis functions for object-centered representations. *Neuron, 37*(2), 347–359. http://www.ncbi.nlm.nih.gov/pubmed/12546828

den Ouden, H. E. M., Daw, N. D., Fernandez, G., Elshout, J. A., Rijpkema, M., Hoogman, M., . . . Cools, R. (2013). Dissociable effects of dopamine and serotonin on reversal learning. *Neuron, 80*, 1090–1100.

den Ouden, H. E. M., Friston, K. J., Daw, N. D., McIntosh, A. R., & Stephan, K. E. (2009). A dual role for prediction error in associative learning. *Cerebral Cortex, 19*(5), 1175–1185. https://doi.org/10.1093/cercor/bhn161

Dienes, Z. (2008). *Understanding psychology as a science: An introduction to scientific and statistical inference*. Palgrave Macmillan.

Efron, B., & Tibshirani, R. J. (1993). *An introduction to the bootstrap*. Chapman & Hall.

Egner, T., Monti, J. M., & Summerfield, C. (2010). Expectation and surprise determine neural population responses in the ventral visual stream. *Journal of Neuroscience, 30*(49), 16601–16608. https://doi.org/10.1523/JNEUROSCI.2770-10.2010

Elbert, T., Elbert, T., Pantev, C., Wienbruch, C., Rockstroh, B., & Taub, E. (1995). Increased cortical representation of the fingers of the left hand in string players. *Science, 270*(5234), 305–307. https://doi.org/10.1126/science.270.5234.305

Elman, J. L. (1990). Finding structure in time. *Cognitive Science, 14*(2), 179–211. https://doi.org/10.1016/0364-0213(90)90002-E

Elman, J. L., Bates, E. A., Johnson, M. H., Karmiloff-Smith, A., Parisi, D., & Plunkett, K. (1996). *Rethinking innateness: A connectionist perspective on development*. MIT Press.

Ergo, K., De Loof, E., & Verguts, T. (2020). Reward prediction error and declarative memory. *Trends in Cognitive Sciences, 24*(5), 388–397.

Farrell, S., & Lewandowsky, S. (2018). *Computational modeling of cognition and behavior*. Cambridge University Press.

Fodor, J. A., & Pylyshyn, Z. (1988). Connectionism and cognitive architecture: A critical analysis. *Cognition, 28*, 3–71.

Franconeri, S. L., Alvarez, G. A., & Cavanagh, P. (2013). Flexible cognitive resources: competitive content maps for attention and memory. *Trends in Cognitive Sciences, 17*(3), 134–141. https://doi.org/10.1016/j.tics.2013.01.010

Frank, M. J., & Badre, D. (2012). Mechanisms of hierarchical reinforcement learning in corticostriatal circuits 1: Computational analysis. *Cerebral Cortex, 22*(3), 509–526. https://doi.org/10.1093/cercor/bhr114

Frank, M. J., Doll, B. B., Oas-Terpstra, J., & Moreno, F. (2009). Prefrontal and striatal dopaminergic genes predict individual differences in exploration and exploitation. *Nature Neuroscience, 12*(8), 1062–1068. https://doi.org/10.1038/nn.2342

References

Frank, M. J., Woroch, B. S., & Curran, T. (2005). Error-related negativity predicts reinforcement learning and conflict biases. *Neuron, 47*(4), 495–501. https://doi.org/10.1016/j.neuron.2005.06.020

Friston, K. J. (2009). The free-energy principle: A rough guide to the brain? *Trends in Cognitive Sciences, 13*(7), 293–301. https://doi.org/10.1016/j.tics.2009.04.005

Friston, K. J. (2010). The free-energy principle: A unified brain theory? *Nature Reviews. Neuroscience, 11*(2), 127–138. https://doi.org/10.1038/nrn2787

Friston, K. J., Glaser, D. E., Henson, R. N., Kiebel, S., Phillips, C., & Ashburner, J. (2002). Classical and Bayesian inference in neuroimaging: Applications. *NeuroImage, 16*(2), 484–512. https://doi.org/10.1006/nimg.2002.1091

Fritzke, B. (1995). A growing neural gas network learns topologies. In *Advances in Neural Information Processing Systems* (pp. 625–632).

Gehring, W. J., & Willoughby, A. R. (2002). The medial frontal cortex and the rapid processing of monetary gains and losses. *Science, 295*(5563), 2279–2282. https://doi.org/10.1126/science.1066893

Gelman, A., Carlin, J. B., Stern, H. S., & Rubin, D. (1995). *Bayesian data analysis.* Chapman & Hall.

Gergely, G., Bekkering, H., & Kiraly, I. (2002). Rational imitation in preverbal infants. *Nature, 415*(February), 755–756.

Gershman, S. J., Blei, D. M., & Niv, Y. (2010). Context, learning, and extinction. *Psychological Review, 117*(1), 197–209. https://doi.org/10.1037/a0017808

Gershman, S. J., Monfils, M., Norman, K. A., & Niv, Y. (2017). The computational nature of memory modification. *ELife*, 1–41. https://doi.org/10.7554/eLife.23763

Gershman, S. J., & Niv, Y. (2010). Learning latent structure: Carving nature at its joints. *Current Opinion in Neurobiology, 20*(2), 251–256. https://doi.org/10.1016/j.conb.2010.02.008

Gill, P., Murray, W., & Wright, M. H. (1982). *Practical optimization.* Emerald Group Publishing.

Grainger, J. (2008). Cracking the orthographic code: An introduction. *Language and Cognitive Processes, 23*, 1–35.

Greve, A., Cooper, E., Kaula, A., Anderson, M. C., & Henson, R. N. (2017). Does prediction error drive one-shot declarative learning? *Journal of Memory and Language, 94*, 149–165. https://doi.org/10.1016/j.jml.2016.11.001

Griffiths, T. L., Chater, N., Kemp, C., Perfors, A., & Tenenbaum, J. B. (2010). Probabilistic models of cognition: exploring representations and inductive biases. *Trends in Cognitive Sciences, 14*(8), 357–364. https://doi.org/10.1016/j.tics.2010.05.004

Griffiths, T. L., Kemp, C., & Tenenbaum, J. B. (2007). Bayesian models of cognition, 1–49. https://cocosci.princeton.edu/tom/papers/bayeschapter.pdf

Griffiths, T. L., & Tenenbaum, J. B. (2006). Optimal predictions in everyday cognition. *Psychological Science, 17*(9), 767–773. https://doi.org/10.1111/j.1467-9280.2006.01780.x

Grill-Spector, K., Henson, R. N., & Martin, A. (2006). Repetition and the brain: Neural models of stimulus-specific effects. *Trends in Cognitive Sciences, 10*(1), 14–23. https://doi.org/10.1016/j.tics.2005.11.006

Grossberg, S. (1980). How does a brain build a cognitive code? *Psychological Review, 87*(1), 1–51. http://www.ncbi.nlm.nih.gov/pubmed/7375607

Hajcak, G. (2015). The reward positivity: From basic research on reward to a biomarker for depression. *Psychophysiology*, *52*(4), 449–459. https://doi.org/10.1111/psyp.12370

Hardin, G. (1968). The tragedy of the commons. *Science*, *162*(3859), 1243–1248.

Hare, M., & Elman, J. L. (1995). Learning and morphological change. *Cognition*, *56*(1), 61–98. http://www.ncbi.nlm.nih.gov/pubmed/22884809

Haselton, M. G., Bryant, G. A., Wilke, A., Frederick, D. A., Galperin, A., Frankenhuis, W. E., & Moore, T. (2009). Adaptive rationality: An evolutionary perspective on cognitive bias. *Social Cognition*, *27*(5), 733–763. https://doi.org/10.1521/soco.2009.27.5.733

Hassabis, D., Kumaran, D., Summerfield, C., & Botvinick, M. M. (2017). Neuroscience-inspired artificial intelligence. *Neuron*, *95*(2), 245–258. https://doi.org/10.1016/j.neuron.2017.06.011

Hasselmo, M. E. (1999). Neuromodulation: Acetylcholine and memory consolidation. *Trends in Cognitive Sciences*, *3*(9), 351–359. http://www.ncbi.nlm.nih.gov/pubmed/10461198

Hasselmo, M. E., & Eichenbaum, H. B. (2005). Hippocampal mechanisms for the context-dependent retrieval of episodes. *Neural Networks*, *18*(9), 1172–1190. https://doi.org/10.1016/j.neunet.2005.08.007

Hasselmo, M. E., & McClelland, J. L. (1999). Neural models of memory. *Current Opinion in Neurobiology*, *9*(2), 184–188. http://www.ncbi.nlm.nih.gov/pubmed/10322183

Hauser, T. U., Iannaccone, R., Ball, J., Mathys, C. D., Brandeis, D., Walitza, S., & Brem, S. (2014). Role of the medial prefrontal cortex in impaired decision making in juvenile attention-deficit/hyperactivity disorder. *JAMA Psychiatry, 71*(10), 1165–1173. https://doi.org/10.1001/jamapsychiatry.2014.1093

Heilbron, M., & Chait, M. (2018). Great expectations: Is there evidence for predictive coding in auditory cortex? *Neuroscience, 389*, 54–73. https://doi.org/10.1016/j.neuroscience.2017.07.061

Hein, A. M., Carrara, F., Brumley, D. R., Stocker, R., & Levin, S. A. (2016). Natural search algorithms as a bridge between organisms, evolution, and ecology. *Proceedings of the National Academy of Sciences, 113*(34), 9413–9420. https://doi.org/10.1073/pnas.1606195113

Henrichs, J. (2004). Demography and cultural evolution: How adaptive cultural processes can produce maladaptive losses—the Tasmanian case. *American Antiquity, 69*(2), 197–214.

Henson, R. N. (1998). Short-term memory for serial order. *Cognitive Psychology, 36*, 73–137.

Heskes, T. (1999). Energy functions for self-organizing maps. *Neural Computation*, 1–13.

Hills, T. T. (2006). Animal foraging and the evolution of goal-directed cognition. *Cognitive Science, 30*(1), 3–41. https://doi.org/10.1207/s15516709cog0000_50

Hills, T. T., Jones, M. N., & Todd, P. M. (2012). Optimal foraging in semantic memory. *Psychological Review, 119*(2), 431–440. https://doi.org/10.1037/a0027373

Hinton, G. E. (1989). Connectionist learning procedures. *Artificial Intelligence, 40*(1–3), 185–234. https://doi.org/10.1016/0004-3702(89)90049-0

Hinton, G. E. (2002). Training products of experts by minimizing contrastive divergence. *Neural Computation, 1800*, 1771–1800.

Hinton, G. E., & Nowlan, S. J. (1987). How learning can guide evolution. *Complex Systems, 1*, 495–502.

References

Hinton, G. E., & Salakhutdinov, R. R. (2006). Reducing the dimensionality of data with neural networks. *Science, 313*(5786), 504–507.

Hinton, G. E., & Sejnowski, T. J. (1983). Optimal perceptual inference. In *Proceedings of the IEEE Conference on Computer Vision and Pattern Recognition,* June.

Hochreiter, S., & Schmidhuber, J. (1997). Long short-term memory. *Neural Computation, 9*(8), 1735–1780.

Holroyd, C. B., & Coles, M. G. H. (2002). The neural basis of human error processing: Reinforcement learning, dopamine, and the error-related negativity. *Psychological Review, 109*(4), 679–709. https://doi.org/10.1037//0033-295X.109.4.679

Holroyd, C. B., Ribas-Fernandes, J., Shahnazian, D., Silvetti, M., & Verguts, T. (2018). Human midcingulate cortex encodes distributed representations of task progress. *Proceedings of the National Academy of Sciences, 115*(25), 6398–6403.

Holroyd, C. B., & Umemoto, A. (2016). The research domain criteria framework: The case for anterior cingulate cortex. *Neuroscience and Biobehavioral Reviews, 71,* 418–443.

Hopfield, J. J. (1982). Neural networks and physical system with emergent collective properties. *Proceedings of the National Academy of Sciences, 79,* 2554–2558.

Hornik, K. (1991). Approximation capabilities of multilayer feedforward networks. *Neural Networks, 4*(1989), 251–257.

Howard, D., Nickels, L., Coltheart, M., & Cole-Virtue, J. (2006). Cumulative semantic inhibition in picture naming: Experimental and computational studies. *Cognition, 100,* 464–482. https://doi.org/10.1016/j.cognition.2005.02.006

Howard, M. W., & Kahana, M. J. (2002). A distributed representation of temporal context. *Journal of Mathematical Psychology, 46*(3), 269–299. https://doi.org/10.1006/jmps.2001.1388

Howard-Jones, P. A., & Jay, T. (2016). Reward, learning and games. *Current Opinion in Behavioral Sciences, 10,* 1–8. https://doi.org/10.1016/j.cobeha.2016.04.015

Hubel, D. H., & Wiesel, T. N. (1968). Receptive fields and functional architecture of monkey striate cortex. *Journal of Neurophysiology, 195,* 215–243.

Hummel, J. E., & Holyoak, K. J. (2003). A symbolic-connectionist theory of relational inference and generalization. *Psychological Review, 110*(2), 220–264. https://doi.org/10.1037/0033-295X.110.2.220

Hunt, L. T., Behrens, T. E. J., Hosokawa, T., Wallis, J. D., & Kennerley, S. W. (2015). Capturing the temporal evolution of choice across prefrontal cortex. *Elife,* 1–25. https://doi.org/10.7554/eLife.11945

Hutcherson, C. A., Bushong, B., & Rangel, A. (2015). A neurocomputational model of altruistic choice and its implications. *Neuron, 87,* 451–462.

Huys, Q. J. M., Daw, N. D., & Dayan, P. (2015). Depression: A decision-theoretic analysis. *Annual Review of Neuroscience, 38,* 1–23. https://doi.org/10.1146/annurev-neuro-071714-033928

Izard, V., & Dehaene, S. (2008). Calibrating the mental number line. *Cognition, 106*(3), 1221–1247. https://doi.org/10.1016/j.cognition.2007.06.004

Jara-Ettinger, J., Gweon, H., Schulz, L. E., & Tenenbaum, J. B. (2016). The naïve utility calculus: Computational principles underlying commonsense psychology. *Trends in Cognitive Sciences, 20*(8), 589–604. https://doi.org/10.1016/j.tics.2016.05.011

Juechems, K., & Summerfield, C. (2019). Where does value come from? *Trends in Cognitive Sciences, 23*(10), P836–P850. https://doi.org/10.1016/j.tics.2019.07.012

Kahneman, D., & Tversky, A. (1979). Prospect theory. *Econometrica, 47*(2), 263–292.

Kamin, L. J. (1969). Predictability, surprise, attention, and conditioning. In B. A. Campbell & R. M. Church (Eds.), *Punishment and aversive behavior* (pp. 279–296). Appleton Century Crofts.

Kandel, E. R. (2001). The molecular biology of memory storage: A dialogue between genes and synapses. *Science, 294*(5544), 1030–1038.

Kennerley, S. W., Walton, M. E., Behrens, T. E. J., Buckley, M. J., & Rushworth, M. F. S. (2006). Optimal decision making and the anterior cingulate cortex. *Nature Neuroscience, 9*(7), 940–947. https://doi.org/10.1038/nn1724

Keramati, M., & Gutkin, B. (2014). Homeostatic reinforcement learning for integrating reward collection and physiological stability. *ELife*, 1–26. https://doi.org/10.7554/eLife.04811

Kim, W., Pitt, M. A., & Myung, I. J. (2013). How do PDP models learn quasiregularity? *Psychological Review, 120*(4), 903–916. https://doi.org/10.1037/a0034195

Kirby, S. (2001). Spontaneous evolution of linguistic structure—an iterated learning model of the emergence of regularity and irregularity. *IEEE Transactions on Evolutionary Computation, 5*(2), 102–110.

Kirby, S., Griffiths, T., & Smith, K. (2014). Iterated learning and the evolution of language. *Current Opinion in Neurobiology, 28*, 108–114. https://doi.org/10.1016/j.conb.2014.07.014

Kirkpatrick, J., Pascanu, R., Rabinowitz, N., Veness, J., Desjardins, G., Rusu, A. A., . . . Hadsell, R. (2017). Overcoming catastrophic forgetting in neural networks. *Proceedings of the National Academy of Sciences, 114*(13), 3521–3526. https://doi.org/10.1073/pnas.1611835114

Kohonen, T. (2001). *Self-organizing maps*. Springer.

Koide-Majima, N., Nakai, T., & Nishimoto, S. (2020). Distinct dimensions of emotion in the human brain and their representation on the cortical surface. *NeuroImage, 222*(November), 117258.

Kool, W., Gershman, S. J., & Cushman, F. A. (2017). Cost-benefit arbitration between multiple reinforcement-learning systems. *Psychological Science, 28*(9), 1321–1333. https://doi.org/10.1177/0956797617708288

Körding, K. P., & Wolpert, D. M. (2004). Bayesian integration in sensorimotor learning. *Nature, 427*(6971), 244–247. https://doi.org/10.1038/nature02169

Körding, K. P., & Wolpert, D. M. (2006). Bayesian decision theory in sensorimotor control. *Trends in Cognitive Sciences, 10*(7), 319–326. https://doi.org/10.1016/j.tics.2006.05.003

Kornysheva, K., Bush, D., Meyer, S. S., Sadnicka, A., Barnes, G., & Burgess, N. (2019). Neural competitive queuing of ordinal structure article neural competitive queuing of ordinal structure underlies skilled sequential action. *Neuron, 101*(6), 1166–1180.

Krajbich, I., Bartling, B., Hare, T., & Fehr, E. (2015). Rethinking fast and slow based on a critique of reaction-time reverse inference. *Nature Communications, 6*(7455), 1–9. https://doi.org/10.1038/ncomms8455

Kriegeskorte, N., Mur, M., & Bandettini, P. (2008). Representational similarity analysis—connecting the branches of systems neuroscience. *Frontiers in Systems Neuroscience, 2*(November), 1–28. https://doi.org/10.3389/neuro.06.004.2008

Kriete, T., Noelle, D. C., Cohen, J. D., & O'Reilly, R. C. (2013). Indirection and symbol-like processing in the prefrontal cortex and basal ganglia. *Proceedings of the National Academy of Sciences*, *110*(41), 16390–16395. https://doi.org/10.1073/pnas.1303547110/-/DCSupplemental.www.pnas.org/cgi/doi/10.1073/pnas.1303547110

Kruschke, J. K. (1992). Alcove: An exemplar-based connectionist model of category learning. *Psychological Review*, *99*(1), 22–44.

Kruschke, J. K. (2006). Locally Bayesian learning with applications to retrospective revaluation and highlighting. *Psychological Review*, *113*(4), 677–699. https://doi.org/10.1037/0033-295X.113.4.677

Kruschke, J. K. (2015). *Doing Bayesian data analysis*. Academic Press.

Kruschke, J. K., & Blair, N. J. (2000). Blocking and backward blocking involve learned inattention. *Psychonomic Bulletin and Review*, *7*(4), 636–645.

Kubilius, J., Bracci, S., & Op De Beeck, H. P. (2016). Deep neural networks as a computational model for human shape sensitivity. *PLoS Computational Biology*, 1–26. https://doi.org/10.1371/journal.pcbi.1004896

Lake, B. M., Ullman, T. D., Tenenbaum, J. B., & Gershman, S. J. (2017). Building machines that learn and think like people. *Behavioral and Brain Sciences*, 1–55. arXiv:1604.00289

LeCun, Y., Bengio, Y., & Hinton, G. E. (2015). Deep learning. *Nature*, *521*(7553), 436–444. https://doi.org/10.1038/nature14539

LeCun, Y., Bottou, L., Bengio, Y., & Haffner, P. (1998). Gradient-based learning applied to document recognition. *Proceedings of the IEEE*, *86*(11), 2278–2323. https://doi.org/10.1109/5.726791

Lee, M. D., & Wagenmakers, E. (2013). *Bayesian models for cognitive science: A practical course*. Cambridge University Press.

Li, P., Farkas, I., & Macwhinney, B. (2004). Early lexical development in a self-organizing neural network, *17*, 1345–1362. https://doi.org/10.1016/j.neunet.2004.07.004

Lieberman, E., Michel, J.-B., Jackson, J., Tang, T., & Nowak, M. A. (2007). Quantifying the evolutionary dynamics of language. *Nature*, *449*(7163), 713–716. https://doi.org/10.1038/nature06137

Lieder, F., & Griffiths, T. L. (2020). Resource-rational analysis: Understanding human cognition as the optimal use of limited computational resources. *Behavioral and Brain Sciences*. https://www.cambridge.org/core/journals/behavioral-and-brain-sciences/article/resourcerational-analysis-understanding-human-cognition-as-the-optimal-use-of-limited-computational-resources/586866D9AD1D1EA7A1EECE217D392F4A

Lillicrap, T. P., Cownden, D., Tweed, D. B., & Akerman, C. J. (2016). Random synaptic feedback weights support error backpropagation for deep learning. *Nature Communications*, *7*, 1–10. https://doi.org/10.1038/ncomms13276

Lillicrap, T. P., Santoro, A., Marris, L., Akerman, C. J., & Hinton, G. E. (2020). Backpropagation and the brain. *Nature Reviews Neuroscience*, *21*(June), 335–346. https://doi.org/10.1038/s41583-020-0277-3

Linsker, R. (1986a). From basic network principles to neural architecture: Emergence of orientation columns. *Proceedings of the National Academy of Sciences of the United States of America*, *83*, 8779–8783.

Linsker, R. (1986b). From basic network principles to neural architecture: Emergence of orientation-selective cells. *Proceedings of the National Academy of Sciences of the United States of America, 83*, 8390–8394.

Linsker, R. (1986c). From basic network principles to neural architecture: Emergence of spatial-opponent cells. *Proceedings of the National Academy of Sciences, 83*, 7508–7512.

Ljungberg, T., Apicella, P., & Schultz, W. (1992). Responses of monkey dopamine neurons during learning of behavioral reactions. *Journal of Neurophysiology, 67*(1), 145–163. http://www.ncbi.nlm.nih.gov/pubmed/1552316

Ma, W. J., Husain, M., & Bays, P. M. (2014). Changing concepts of working memory. *Nature Neuroscience, 17*(3), 347–356. https://doi.org/10.1038/nn.3655

Maes, E., Boddez, Y., Alfei, J. M., Krypotos, A.-M., D'Hooge, R., De Houwer, J., & Beckers, T. (2016). The elusive nature of the blocking effect: 15 failures to replicate. *Journal of Experimental Psychology: General, 145*(9), 49–71. https://doi.org/10.1037/xge0000200

Mahmoodi, A., Bang, D., Olsen, K., Aimee, Y., Shi, Z., & Broberg, K. (2015). Equality bias impairs collective decision-making across cultures. *Proceedings of the National Academy of Sciences, 112*(12), 3835–3840. https://doi.org/10.1073/pnas.1421692112

Maia, T. V, & Frank, M. J. (2011). From reinforcement learning models to psychiatric and neurological disorders. *Nature Neuroscience, 14*(2), 154–162. https://doi.org/10.1038/nn.2723

Mamassian, P., Landy, M., & Maloney, L. T. (2001). Bayesian modelling of visual perception. In R. P. N. Rao, B. A. Olshausen, & M. S. Lewicki (Eds.), *Probabilistic models of the brain: Perception and neural function* (pp. 13–36). MIT Press.

Mansfield, E. L., Karayanidis, F., Jamadar, S., Heathcote, A., & Forstmann, B. U. (2011). Adjustments of response threshold during task switching: A model-based functional magnetic resonance imaging study. *Journal of Neuroscience, 31*(41), 14688–14692. https://doi.org/10.1523/jneurosci.2390-11.2011

Marcus, G. F. (2006). *The algebraic mind: Integrating connectionism and cognitive science*. MIT Press.

Marcus, G. F. (2018). Deep learning: A critical appraisal. *Arxiv*, 1–27. arXiv:1801.00631

Marcus, G., & Davis, E. (2019). *Rebooting AI: Building artificial intelligence we can trust*. Vintage Books.

Maris, E., & Oostenveld, R. (2007). Nonparametric statistical testing of EEG- and MEG-data. *Journal of Neuroscience Methods, 164*(1), 177–190. https://doi.org/10.1016/j.jneumeth.2007.03.024

Marr, D. (1982). *Vision: A computational investigation into the human representation and processing of visual information*. W. H. Freeman.

Mathys, C. D., Daunizeau, J., Friston, K. J., & Stephan, K. E. (2011). A Bayesian foundation for individual learning under uncertainty. *Frontiers in Human Neuroscience, 5*(May), 39. https://doi.org/10.3389/fnhum.2011.00039

Mattar, M. G., & Daw, N. D. (2018). Prioritized memory access explains planning and hippocampal replay. *Nature Neuroscience, 21*(November). https://doi.org/10.1038/s41593-018-0232-z

Maxwell, S. E., Delaney, H. D., & Kelley, K. (2004). *Designing experiments and analyzing data: A model comparison perspective*. Routledge.

Mayor, J., & Plunkett, K. (2010). A neurocomputational account of taxonomic responding and fast mapping in early word learning. *Psychological Review, 117*(1), 1–31. https://doi.org/10.1037/a0018130

References   **231**

McClelland, J. L., & Elman, J. L. (1986). The TRACE model of speech perception. *Cognitive Psychology, 18*, 1–86. https://doi.org/10.1016/0010-0285(86)90015-0

McClelland, J. L., McNaughton, B. L., & O'Reilly, R. C. (1995). Why there are complementary learning systems in the hippocampus and neocortex. *Psychological Review, 102*, 419–457.

McClelland, J. L., & Patterson, K. (2002). "Words or rules" cannot exploit the regularity in exceptions. *Trends in Cognitive Sciences, 6*(11), 464–465.

McClelland, J. L., & Rumelhart, D. E. (1981). An interactive activation model of context effects in letter perception: Part 1. An account of basic findings. *Psychological Review, 88*(5), 375–407.

McCloskey, M. (1991). Networks and theories: The place of connectionism in cognitive science. *Psychological Science, 2*(6), 387–395.

McCloskey, M., & Cohen, N. J. (1989). Catastrophic interference in connectionist networks: The sequential learning problem. *Psychology of Learning & Motivation, 24*, 109–165.

McLeod, P., Plunkett, K., & Rolls, E. T. (1998). *Introduction to connectionist modeling of cognitive processes.* Oxford University Press.

McRae, K. (2004). Semantic memory: Some insights from feature-based connectionist attractor networks. In B. H. Ross (Ed.), *The psychology of learning and motivation* (Vol. 45). Academic Press.

Mercier, H., & Sperber, D. (2011). Why do humans reason? Arguments for an argumentative theory. *Behavioral and Brain Sciences, 34*(2), 57–74; discussion 74–111. https://doi.org/10.1017/S0140525X10000968

Mesoudi, A. (2011). *Cultural evolution: How Darwinian theory can explain human culture and synthesize the social sciences.* University of Chicago Press.

Mestdagh, M., Verdonck, S., Meers, R., Loossens, T., & Tuerlinckx, F. (2019). Prepaid parameter estimation without likelihoods. *PLoS Computational Biology, 15(9)*, 1–42. https://doi.org/10.1371/journal.pcbi.1007181

Meyniel, F., Schlunegger, D., & Dehaene, S. (2015). The sense of confidence during probabilistic learning: A normative account. *PLoS Computational Biology*, 1–25. https://doi.org/10.1371/journal.pcbi.1004305

Miller, G. A. (1956). The magical number 7, plus or minus 2: Some limits on our capacity for processing information. *Psychological Review, 63*, 81–97.

Miller, G. A., Galanter, E., & Pribram, K. H. (1960). *Plans and the structure of behavior.* Henry Holt and Co. https://doi.org/10.1037/10039-000

Millidge, B., Tschantz, A., & Buckley, C. L. (2020). Whence the expected free energy? *ArXiv.*

Milner, B., & Scoville, W. B. (1957). Loss of recent memory after bilateral hippocampal lesions. *Journal of Neurology, Neurosurgery and Psychiatry, 20*(11), 11. https://doi.org/10.1136/jnnp-2015-311092

Minda, J. P., & Smith, J. D. (2011). Prototype models of categorization: Basic formulation, predictions, and limitations. In E. M. Pothos & A. J. Wills (Eds.), *Formal approaches in categorization* (pp. 40–64). Cambridge University Press.

Minsky, M., & Papert, S. (1969). *Perceptrons: An introduction to computational geometry.* MIT Press.

Miynarski, W., Hledik, M., Sokolowski, T. R., & Tkacik, G. (2020). Statistical analysis and optimality of biological systems. *bioRxxiv*, 1–13. https://www.biorxiv.org/content/10.1101/848374v2

Mnih, V., Kavukcuoglu, K., Silver, D., Rusu, A. A., Veness, J., Bellemare, M. G., . . . Hassabis, D. (2015). Human-level control through deep reinforcement learning. *Nature*, *518*(7540), 529–533. https://doi.org/10.1038/nature14236

Momennejad, I., Otto, A. R., Daw, N. D., & Norman, K. A. (2018). Offline replay supports planning in human reinforcement learning. *ELife*, 1–25.

Montague, P. R., Dayan, P., & Sejnowski, T. J. (1996). A framework for mesencephalic dopamine systems based on predictive Hebbian learning. *Journal of Neuroscience*, *16*(5), 1936–1947.

Mood, A. M., Graybill, F. A., & Boes, D. C. (1973). *Introduction to the theory of statistics*. McGraw-Hill.

Morris, R. G. M., Garrud, P., Rawlins, J. N. P., & O'Keefe, J. (1982). Place navigation impaired in rats with hippocampal lesions. *Nature*, 297, 681–683.

Moyer, R. S., & Landauer, T. K. (1967). Time required for judgements of numerical inequality. *Nature*, *215*, 1519–1520.

Mulder, M. J., Wagenmakers, E.-J., Ratcliff, R., Boekel, W., & Forstmann, B. U. (2012). Bias in the brain: A diffusion model analysis of prior probability and potential payoff. *Journal of Neuroscience*, *32*(7), 2335–2343. https://doi.org/10.1523/JNEUROSCI.4156-11.2012

Mullen, K. M., Ardia, D., Gil, D. L., Windover, D., & Cline, J. (2009). DEoptim: An R package for global optimization by differential evolution. MPRA Paper No. 27878. http://mpra.ub.uni-muenchen.de/27878/2/MPRA_paper_27878.pdf

Murphy, G. L., & Medin, D. L. (1985). The role of theories in conceptual coherence. *Psychological Review*, *92*(3), 289–316. https://doi.org/10.1037//0033-295X.92.3.289

Murphy, K. (2012). *Machine learning: A Probabilistic Perspective*. MIT Press.

Murty, V. P., DuBrow, S., & Davachi, L. (2015). The simple act of choosing influences declarative memory. *Journal of Neuroscience*, *35*(16), 6255–6264. https://doi.org/10.1523/JNEUROSCI.4181-14.2015

Muthukrishna, M., & Henrichs, J. (2019). A problem in theory. *Nature Human Behaviour*, *3*, 221–229.

Ness, T., & Meltzer-asscher, A. (2021). Love thy neighbor: Facilitation and inhibition in the competition between parallel predictions. *Cognition*, *207*(September 2020).

Nieder, A. (2017). Number faculty is rooted in our biological heritage. *Trends in Cognitive Sciences*, *21*(6), 403–404. https://doi.org/10.1016/j.tics.2017.03.014

Nieder, A., Freedman, D. J., & Miller, E. K. (2002). Representation of the quantity of visual items in the primate prefrontal cortex. *Science*, *297*(5587), 1708–1711. https://doi.org/10.1126/science.1072493

Nikkilä, J., Törönen, P., Kaski, S., Venna, J., Castrén, E., & Wong, G. (2002). Analysis and visualization of gene expression data using self-organizing maps. *Neural Networks*, *15*(8–9), 953–966. https://doi.org/10.1016/S0893-6080(02)00070-9

Niv, Y., Daw, N. D., Joel, D., & Dayan, P. (2007). Tonic dopamine: Opportunity costs and the control of response vigor. *Psychopharmacology*, *191*(3), 507–520. https://doi.org/10.1007/s00213-006-0502-4

Nosofsky, R. M., Sanders, C. A., & McDaniel, M. A. (2018). A formal psychological model of classification applied to natural-science category learning. *Current Directions in Psychological Science*, *27*(2), 129–135. https://doi.org/10.1177/0963721417740954

Nowak, M. A. (2006). Five rules for the evolution of cooperation. *Science, 314*(5805), 1560–1563. https://doi.org/10.1126/science.1133755

Nowak, M. A., & Sigmund, K. (1993). A strategy of win-stay, lose-shift that outperforms tit-for-tat in the Prisoner's Dilemma game. *Nature, 364*, 56–58.

Oberauer, K., & Lewandowsky, S. (2019). Addressing the theory crisis in psychology. *Psychonomic Bulletin and Review, 26*, 1596–1618.

Oberauer, K., & Lin, H. (2017). An Interference Model of Visual Working Memory. *Psychological Review, 124*(1), 21–59.

O'Doherty, J. P., Dayan, P., Schultz, J., Deichmann, R., Friston, K. J., & Dolan, R. J. (2004). Dissociable roles of ventral and dorsal striatum in instrumental conditioning. *Science, 304*(5669), 452–454. https://doi.org/10.1126/science.1094285

Oja, E. (1982). A simplified neuron model as a principal component analyzer. *Journal of Mathematical Biology, 1*, 267–273.

Oliveira, F. T. P., McDonald, J. J., & Goodman, D. (2007). Performance monitoring in the anterior cingulate is not all error related: expectancy deviation and the representation of action-outcome associations. *Journal of Cognitive Neuroscience, 19*(12), 1994–2004. https://doi.org/10.1162/jocn.2007.19.12.1994

Open science collaboration. (2015). Estimating the reproducibility of psychological science. *Science*, 943–950.

Oppenheim, G. M., Dell, G. S., & Schwartz, M. F. (2010). The dark side of incremental learning: A model of cumulative semantic interference during lexical access in speech production. *Cognition, 114*(2), 227–252. https://doi.org/10.1016/j.cognition.2009.09.007

O'Reilly, R. C. (2006). Models of high-level cognition. *Science*, October, 91–94.

O'Reilly, R. C., & Frank, M. J. (2006). Making working memory work: A computational model of learning in the prefrontal cortex and basal ganglia. *Neural Computation, 18*(2), 283–328. https://doi.org/10.1162/089976606775093909

O'Reilly, R. C., & Norman, K. A. (2002). Hippocampal and neocortical contributions to memory: advances in the complementary learning systems framework. *Trends in Cognitive Sciences, 6*(12), 505–510. http://www.ncbi.nlm.nih.gov/pubmed/12475710

Otto, A. R., Gershman, S. J., Markman, A. B., & Daw, N. D. (2013a). The curse of planning: Dissecting multiple reinforcement-learning systems by taxing the central executive. *Psychological Science, 24*(5), 751–761. https://doi.org/10.1177/0956797612463080

Otto, A. R., Raio, C. M., Chiang, A., Phelps, E. A., & Daw, N. D. (2013b). Working-memory capacity protects model-based learning from stress. *Proceedings of the National Academy of Sciences, 110*(52). https://doi.org/10.1073/pnas.1312011110

Oudeyer, P. (2005). The self-organization of speech sounds, *Journal of Theoretical Biology, 233*(3), 435–449. https://doi.org/10.1016/j.jtbi.2004.10.025

Page, M. P. A., & Norris, D. (1998). The primacy model: A new model of immediate serial recall. *Psychological Review, 105*(4), 761–781. https://doi.org/10.1037/0033-295X.105.4.761-781

Palminteri, S., Khamassi, M., Joffily, M., & Coricelli, G. (2015). Contextual modulation of value signals in reward and punishment learning. *Nature Communications, 6*, 8096. https://doi.org/10.1038/ncomms9096

Palminteri, S., Wyart, V., & Koechlin, E. (2017). The importance of falsification in computational cognitive modeling. *Trends in Cognitive Sciences, 21*(6), 425–433. https://doi.org/10.1016/j.tics.2017.03.011

Peirce, J. W. (2007). PsychoPy—psychophysics software in Python. *Journal of Neuroscience Methods, 162*(1–2), 8–13. https://doi.org/10.1016/j.jneumeth.2006.11.017

Peters, J., & Schaal, S. (2008). Reinforcement learning of motor skills with policy gradients. *Neural Networks, 21*, 682–697. https://doi.org/10.1016/j.neunet.2008.02.003

Piazza, M., Pinel, P., Le Bihan, D., & Dehaene, S. (2007). A magnitude code common to numerosities and number symbols in human intraparietal cortex. *Neuron, 53*(2), 293–305. https://doi.org/10.1016/j.neuron.2006.11.022

Pine, A., Sadeh, N., Ben-yakov, A., Dudai, Y., & Mendelsohn, A. (2018). Knowledge acquisition is governed by striatal prediction errors. *Nature Communications, 9*, 1673. https://doi.org/10.1038/s41467-018-03992-5

Pinker, S., & Prince, A. (1988). On language and connectionism: Analysis of a parallel distributed processing model of language processing. *Cognition, 28*, 73–193.

Pinker, S., & Ullman, M. T. (2002). The past and future of the past tense. *Trends in Cognitive Sciences, 6*(11), 456–463.

Piray, P., Dezfouli, A., Heskes, I., Frank, M. J., & Daw, N. D. (2019). Hierarchical Bayesian inference for concurrent model fitting and comparison for group studies. *PLoS Computational Biology, 15*(6), https://doi.org/10.1371/journal.pcbi.1007043

Pitt, M. A, Kim, W., Navarro, D. J., & Myung, I. J. (2006). Global model analysis by parameter space partitioning. *Psychological Review, 113*(1), 57–83. https://doi.org/10.1037/0033-295X.113.1.57

Pitt, M. A., & Myung, I. J. (2002). When a good fit can be bad. *Trends in Cognitive Sciences, 6*(10), 421–425. http://www.ncbi.nlm.nih.gov/pubmed/12413575

Pitt, M. A., Myung, I. J., & Zhang, S. (2002). Toward a method of selecting among computational models of cognition. *Psychological Review, 109*(3), 472–491. https://doi.org/10.1037//0033-295X.109.3.472

Plaut, D. C., McClelland, J. L., Seidenberg, M. S., & Patterson, K. (1996). Understanding normal and impaired word reading: Computational principles in quasi-regular domains. *Psychological Review, 103*(1), 56–115. https://doi.org/10.1037/0033-295X.103.1.56

Plunkett, K., & Marchman, V. (1993). From rote learning to system building: Acquiring verb morphology in children and connectionist nets. *Cognition, 48*, 21–69.

Poggio, T., & Bizzi, E. (2004). Generalization in vision and motor control. *Nature, 431*(October), 768–774.

Powell, A., Shennan, S., & Thomas, M. G. (2009). Late Pleistocene demography and the appearance of modern human behavior. *Science, 324*(5932), 1298–1301. https://doi.org/10.1126/science.1170165

Price, G. R. (1970). Selection and covariance. *Nature, 227*, 520–521.

Prince, A., & Smolensky, P. (1997). Optimality: From neural networks to universal grammar. *Science, 1604*. https://doi.org/10.1126/science.275.5306.1604

Punch, W. F., & Enbody, R. (2014). *The practice of computing using Python*. Pearson.

Quax, S., & Van Gerven, M. (2018). Emergent mechanisms of evidence integration in recurrent neural networks. *PloS One, 13*(10), 1–22.

Rand, D. G., & Nowak, M. A. (2013). Human cooperation. *Trends in Cognitive Sciences, 17*(8), 413–425. https://doi.org/10.1016/j.tics.2013.06.003

Rao, R. P., & Ballard, D. H. (1999). Predictive coding in the visual cortex: A functional interpretation of some extra-classical receptive-field effects. *Nature Neuroscience, 2*(1), 79–87. https://doi.org/10.1038/4580

Ratcliff, R. (1978). A theory of memory retrieval. *Psychological Review, 85*(2), 59–108. https://doi.org/10.1037/0033-295X.85.2.59

Ratcliff, R., Gomez, P., Thapar, A., & McKoon, G. (2004). A diffusion model analysis of the effects of aging in the lexical-decision task. *Psychology and Aging, 19*(2), 278–289. https://doi.org/10.1037/0882-7974.19.2.278

Ratcliff, R., & Rouder, J. N. (1998). Modeling response times for two-choice decisions. *Psychological Science, 9*(5), 347–356.

Ratcliff, R., & Tuerlinckx, F. (2002). Estimating parameters of the diffusion model: Approaches to dealing with contaminant reaction times and parameter variability. *Psychonomic Bulletin & Review, 9*(3), 438–481.

Ravichandiran, S. (2018). *Reinforcement learning with Python*. Packt.

Raworth, K. (2017). *Doughnut economics*. Penguin.

Reali, F., Chater, N., & Christiansen, M. H. (2018). Simpler grammar, larger vocabulary: How population size affects language. *Proceedings of the Royal Society B: Biological Sciences, 285*(1871). https://royalsocietypublishing.org/doi/10.1098/rspb.2017.2586

Reicher, G. M. (1969). Perceptual recognition as a function of meaningful stimulus material. *Journal of Experimental Psychology, 2*, 275–280.

Ren, S., He, K., Girshick, R., & Sun, J. (2017). Faster R-CNN: Towards real-time object detection with region proposal networks. *IEEE Transactions on Pattern Analysis and Machine Intellligence, 39*(6), 1137–1149.

Rescorla, R. A., & Wagner, A. (1972). A theory of Pavlovian conditioning: Variations in the effectiveness of reinforcement and nonreinforcement. In A. Black & W. Prokasy (Eds.), *Classical conditioning II: Current research and theory* (pp. 64–99). Appleton Century Crofts.

Reynolds, J. N. J., Hyland, B. I., & Wickens, J. R. (2001). A cellular mechanism of reward-related learning. *Nature, 413*(6851), 67–70. https://doi.org/10.1038/35092560

Roediger, H. L., & Crowder, R. G. (1976). A serial position effect in recall of United States presidents. *Bulletin of the Psychonomic Society, 8*(4), 275–278.

Roelfsema, P. R., & van Ooyen, A. (2005). Attention-gated reinforcement learning of internal representations for classification. *Neural Computation, 17*(10), 2176–2214. https://doi.org/10.1162/0899766054615699

Roelfsema, P. R., van Ooyen, A., & Watanabe, T. (2010). Perceptual learning rules based on reinforcers and attention. *Trends in Cognitive Sciences, 14*(2), 64–71. https://doi.org/10.1016/j.tics.2009.11.005

Rogers, T. T., Lambon Ralph, M. A, Garrard, P., Bozeat, S., McClelland, J. L., Hodges, J. R., & Patterson, K. (2004). Structure and deterioration of semantic memory: A neuropsychological and computational investigation. *Psychological Review, 111*(1), 205–235. https://doi.org/10.1037/0033-295X.111.1.205

Rogers, T. T., & McClelland, J. L. (2004). *Semantic cognition: A parallel distributed processing approach.* MIT Press.

Roggeman, C., Santens, S., Fias, W., & Verguts, T. (2011). Stages of nonsymbolic number processing in occipitoparietal cortex disentangled by fMRI adaptation. *Journal of Neuroscience, 31*(19), 7168–7173. https://doi.org/10.1523/JNEUROSCI.4503-10.2011

Rojas, R. (1996). *Neural networks: A systematic introduction.* Springer.

Rolls, E. T., Loh, M., Deco, G., & Winterer, G. (2008). Computational models of schizophrenia and dopamine modulation in the prefrontal cortex. *Nature Reviews Neuroscience, 9*(9), 696–709. https://doi.org/10.1038/nrn2462

Rombouts, J. O., Bohte, S. M., & Roelfsema, P. R. (2015). How attention can create synaptic tags for the learning of working memories in sequential tasks. *PLoS Computational Biology, 11*(3), 1–34. https://doi.org/10.1371/journal.pcbi.1004060

Rosenblatt, F. (1958). The perceptron: A probabilistic model for information storage and organization in the brain. *Psychological Review, 65*(6), 386–408.

Rouhani, N., Norman, K. A., & Niv, Y. (2018). Dissociable effects of surprising rewards on learning and memory. *Journal of Experimental Psychology: Learning, Memory, and Cognition, 44*(9), 1430–1443.

Rumelhart, D. E., & McClelland, J. L. (1982). An interactive activation model of context effects in letter perception: Part 2. The contextual enhancement effect and some tests and extensions of the model. *Psychological Review, 89*(I), 60–94.

Rumelhart, D. E., & McClelland, J. L. (1986). On learning the past tenses of English verbs. In J. L. McClelland, D. E. Rumelhart, & the PDP Research Group (Eds.), *Parallel distributed processing; Explorations in the microstructure of cognition* (pp. 216–271). *Volume 2: Psychological and biological models.* MIT Press.

Rumelhart, D. E., Williams, & Hinton, G. E. (1986). Learning internal representations by error propagation. In D. Rumelhart, J. L. McClelland, & The PDP Research Group (Eds.), *Parallel Distributed Processing; Explorations in the microstructure of cognition* (pp. 318–362). *Volume 1: Foundations.* MIT Press.

Rumelhart, D. E., McClelland, J. L., & The PDP Research Group. (1986b). *Parallel distributed processing: Explorations in the microstructure of cognition. Volume 1: Foundations.* MIT Press.

Sahani, M. (1999). *Latent variable models for neural data analysis.* California Institute of Technology. http://www.gatsby.ucl.ac.uk/~maneesh/thesis/thesis.double.pdf

Sambrook, T. D., & Goslin, J. (2015). A neural reward prediction error revealed by a meta-analysis of ERPs using great grand averages. *Psychological Bulletin, 141*(1), 213–235. https://doi.org/10.1037/bul0000006

Sanger, T. D. (1989). Optimal unsupervised learning in a single-layer linear feedforward neural network. *Neural Networks, 2*, 459–473.

Sara, S. J., & Bouret, S. (2012). Orienting and reorienting: The locus coeruleus mediates cognition through arousal. *Neuron, 76*, 130–141.

References

Saxe, A. M., McClelland, J. L., & Ganguli, S. (2019). A mathematical theory of semantic development in deep neural networks. *Proceedings of the National Academy of Sciences, 116*(23), 11537–11546. https://doi.org/10.1073/pnas.1820226116

Schmidhuber, J. (2014). Deep learning in neural networks: An overview. *Neural and Evolutionary Computing; Learning.* http://arxiv.org/abs/1404.7828

Schultz, W., Dayan, P., & Montague, P. R. (1997). A neural substrate of prediction and reward. *Science, 275*(5306), 1593–1599. https://doi.org/10.1126/science.275.5306.1593

Schulz, E., Tenenbaum, J. B., Duvenaud, D., Speekenbrink, M., & Gershman, S. J. (2017). Compositional inductive biases in function learning. *Cognitive Psychology, 99*(December), 44–79. https://doi.org/10.1016/j.cogpsych.2017.11.002

Schulz, E., Wu, C. M., Ruggeri, A., & Meder, B. (2019). Searching for rewards like a child means less generalization and more directed exploration. *Psychological Science, 30*(11), 1561–1572.

Sejnowski, T. J. (2020). The unreasonable effectiveness of deep learning in artificial intelligence. *Proceedings of the National Academy of Sciences, 117*(48), 30033–30038. https://doi.org/10.1073/pnas.1907373117

Sevenster, D., Beckers, T., & Kindt, M. (2013). Prediction error governs pharmacologically induced amnesia for learned fear. *Science, 339*, 830–833. https://doi.org/10.1126/science.1231357

Seymour, B., O'Doherty, J. P., Dayan, P., Koltzenburg, M., Jones, A. K., Dolan, R. J., . . . Frackowiak, R. S. (2004). Temporal difference models describe higher-order learning in humans. *Nature, 429*(June), 664–667. https://doi.org/10.1038/nature02636.1.

Shahnazian, D., & Holroyd, C. B. (2017). Distributed representations of action sequences in anterior cingulate cortex: A recurrent neural network approach. *Psychonomic Bulletin & Review, 25*, 302–331. https://doi.org/10.3758/s13423-017-1280-1

Shenhav, A., Botvinick, M. M., & Cohen, J. D. (2013). The expected value of control: An integrative theory of anterior cingulate cortex function. *Neuron, 79*(2), 217–240. https://doi.org/10.1016/j.neuron.2013.07.007

Shennan, S. (2002). *Genes, memes, and human history.* Thames & Hudson.

Silver, D., Huang, A., Maddison, C. J., Guez, A., Sifre, L., van den Driessche, G., . . . Hassabis, D. (2016). Mastering the game of Go with deep neural networks and tree search. *Nature, 529*(7587), 484–489. https://doi.org/10.1038/nature16961

Silver, M. A., & Kastner, S. (2009). Topographic maps in human frontal and parietal cortex. *Trends in Cognitive Sciences, 13*(11), 488–495. https://doi.org/10.1016/j.tics.2009.08.005

Silvetti, M., Alexander, W. H., Verguts, T., & Brown, J. W. (2014). From conflict management to reward-based decision making: Actors and critics in primate medial frontal cortex. *Neuroscience and Biobehavioral Reviews, 46*, 44–57.

Silvetti, M., Nuñez Castellar, E., Roger, C., & Verguts, T. (2014). Reward expectation and prediction error in human medial frontal cortex: An EEG study. *NeuroImage, 84*, 376–382. https://doi.org/10.1016/j.neuroimage.2013.08.058

Silvetti, M., Vassena, E., Abrahamse, E. L., & Verguts, T. (2018). Dorsal anterior cingulate-brainstem ensemble as a reinforcement meta-learner. *PLOS Computational Biology.* https://doi.org/https://doi.org/10.1371/journal.pcbi.1006370

Skinner, B. (1938). *The behavior of organisms: An experimental analysis.* Appleton-Century-Croft.

Solway, A., & Botvinick, M. M. (2015). Evidence integration in model-based tree search. *Proceedings of the National Academy of Sciences, 112*(37), 201505483. https://doi.org/10.1073/pnas.1505483112

Squire, L. R., & Kandel, E. R. (1999). *Memory: From mind to molecules.* W. H. Freeman.

Steegen, S., Tuerlinckx, F., & Vanpaemel, W. (2017). Using parameter space partitioning to evaluate a model's qualitative fit. *Psychonomic Bulletin & Review, 24,* 617–631. https://doi.org/10.3758/s13423-016-1123-5

Steels, L. (2011). Modeling the cultural evolution of language. *Physics of Life Reviews, 8,* 339–356. https://doi.org/10.1016/j.plrev.2011.10.014

Steinberg, E. E., Keiflin, R., Boivin, J. R., Witten, I. B., Deisseroth, K., & Janak, P. H. (2013). A causal link between prediction errors, dopamine neurons and learning. *Nature Neuroscience, 16*(7), 966–973. https://doi.org/10.1038/nn.3413

Stoianov, I., & Zorzi, M. (2012). Emergence of a "visual number sense" in hierarchical generative models. *Nature Neuroscience, 15*(2), 194–196. https://doi.org/10.1038/nn.2996

Stojic, H., Orquin, J. L., Dayan, P., Dolan, R. J., & Speekenbrink, M. (2020). Uncertainty in learning, choice, and visual fixation. *Proceedings of the National Academy of Sciences, 117*(6), 3291–3300. https://doi.org/10.1073/pnas.1911348117

Sutton, R. S., & Barto, A. G. (2018). *Reinforcement learning: An introduction.* 2nd ed. MIT Press.

Tang, Y. P., Shimizu, E., Dube, G. R., Rampon, C., Kerchner, G. A, Zhuo, M., . . . Tsien, J. Z. (1999). Genetic enhancement of learning and memory in mice. *Nature, 401*(6748), 63–69. https://doi.org/10.1038/43432

Tanner, M. A. (1996). *Tools for statistical inference.* Springer.

Ten Berge, J. M. F. (2005). *Least squares optimization in multivariate analysis* (Vol. 34). DSWO Press.

Tesauro, G. (1995). Temporal difference learning and TD-Gammon. *Communications of the ACM, 38*(3). https://dl.acm.org/doi/10.1145/203330.203343

Testolin, A., Stoianov, I., De Filippo De Grazia, M., & Zorzi, M. (2013). Deep unsupervised learning on a desktop PC: A primer for cognitive scientists. *Frontiers in Psychology, 4*(May). https://doi.org/10.3389/fpsyg.2013.00251

Testolin, A., Stoianov, I., & Zorzi, M. (2017). Letter perception emerges from unsupervised deep learning and recycling of natural image features. *Nature Human Behaviour, 1,* 657–664. https://doi.org/10.1038/s41562-017-0186-2

Thorndike, E. L. (1901). The mental life of the monkeys. *Psychological Review: Monograph Supplements, 3*(5), i–57. https://doi.org/10.1037/h0092994

Tolman, E. C. (1948). Cognitive maps in rats and men. *Psychological Review, 55,* 189–208. https://doi.org/10.1037/h0061626

Treisman, A. (1996). The binding problem. *Current Opinion in Neurobiology, 6*(2), 171–178. http://www.ncbi.nlm.nih.gov/pubmed/10677031

Treisman, A. (2006). How the deployment of attention determines what we see. *Visual cognition* (Vol. 14). https://doi.org/10.1080/13506280500195250

References

Tuerlinckx, F., Maris, E., Ratcliff, R., & De Boeck, P. (2001). A comparison of four methods for simulating the diffusion process. *Behavior Research Methods, Instruments, & Computers, 33*(4), 443–456. https://doi.org/10.3758/BF03195402

Turner, B. M., Forstmann, B. U., Wagenmakers, E.-J., Brown, S. D., Sederberg, P. B., & Steyvers, M. (2013). A Bayesian framework for simultaneously modeling neural and behavioral data. *NeuroImage, 72*, 193–206. https://doi.org/10.1016/j.neuroimage.2013.01.048

Turner, B. M., & Van Zandt, T. (2018). Approximating Bayesian inference through model simulation. *Trends in Cognitive Sciences, 22*(9), 826–840. https://doi.org/10.1016/j.tics.2018.06.003

Usher, M., & McClelland, J. L. (2001). The time course of perceptual choice: The leaky, competing accumulator model. *Psychological Review, 108*(3), 550–592.

van den Berg, R., Awh, E., & Ma, W. J. (2014). Factorial comparison of working memory models. *Psychological Review, 121*(1), 124–149. https://doi.org/10.1037/a0035234

van den Heuvel, M., Sporns, O., Collin, G., Scheewe, T., Mandi, R. C. W., Cahn, W., . . . Kahn, R. S. (2013). Abnormal rich club organization and functional brain dynamics in schizophrenia. *JAMA Psychiatry, 70*(8), 783–792. https://doi.org/10.1001/jamapsychiatry.2013.1328

Vandekerckhove, J., & Tuerlinckx, F. (2008). Diffusion model analysis with MATLAB: A DMAT primer. *Behavior Research Methods, 40*(1), 61–72. https://doi.org/10.3758/BRM.40.1.61

Vanpaemel, W., & Lee, M. D. (2012). Using priors to formalize theory: Optimal attention and the generalized context model. *Psychonomic Bulletin and Review, 19*, 1047–1056. https://doi.org/10.3758/s13423-012-0300-4

van Rooij, I., & Baggio, G. (2020). Theory development requires an epistemological sea change. *Psychological Inquiry, 31*(4), 321–325. https://doi.org/10.1080/1047840X.2020.1853477

Verbeke, P., & Verguts, T. (2019). Learning to synchronize: How biological agents can couple neural task modules for dealing with the stability-plasticity dilemma. *PLoS Computational Biology.* https://journals.plos.org/ploscompbiol/article?id=10.1371/journal.pcbi.1006604

Verdonck, S., & Tuerlinckx, F. (2014). The Ising decision maker: A binary stochastic network for choice response time. *Psychological Review, 121*(3), 422–462. https://doi.org/10.1037/a0037012

Verguts, T., & Fias, W. (2004). Representation of number in animals and humans: A neural model. *Journal of Cognitive Neuroscience, 16*(9), 1493–1504. https://doi.org/10.1162/0898929042568497

Verguts, T., Fias, W., & Stevens, M. (2005). A model of exact small-number representation. *Psychonomic Bulletin & Review, 12*(1), 66–80. http://www.ncbi.nlm.nih.gov/pubmed/15945201

Verguts, T., & Notebaert, W. (2008). Hebbian learning of cognitive control: dealing with specific and nonspecific adaptation. *Psychological Review, 115*(2), 518–525. https://doi.org/10.1037/0033-295X.115.2.518

Verguts, T., Vassena, E., & Silvetti, M. (2015). Adaptive effort investment in cognitive and physical tasks: A neurocomputational model. *Frontiers in Behavioral Neuroscience, 9*, 1–17. https://doi.org/https://doi.org/10.3389/fnbeh.2015.00057

Vermunt, J. K., & Magidson, J. (2004). Latent class analysis. In *Sage encyclopedia of social science research methods.*

Vinckier, F., Dehaene, S., Jobert, A., Dubus, J. P., Sigman, M., & Cohen, L. (2007). Hierarchical coding of letter strings in the ventral stream: Dissecting the inner organization of the visual word-form system. *Neuron, 55*(1), 143–156. https://doi.org/10.1016/j.neuron.2007.05.031

von der Malsburg, C. (1995). Binding in models of perception and brain function. *Current Opinion in Neurobiology, 5*, 520–526.

von Neumann, J., & Morgenstern, O. (1944). *Theory of games and economic behavior.* Princeton University Press.

Wagenmakers, E.-J., & Farrell, S. (2004). AIC model selection using Akaike weights. *Psychonomic Bulletin & Review, 11*(1), 192–196.

Wagenmakers, E.-J., van der Maas, H. L. J., & Grasman, R. P. P. P. (2007). An EZ-diffusion model for response time and accuracy. *Psychonomic Bulletin & Review, 14*(1), 3–22. http://www.ncbi.nlm.nih.gov/pubmed/17546727

Wang, J. X., Kurth-nelson, Z., Kumaran, D., Tirumala, D., Soyer, H., Leibo, J. Z., . . . Botvinick, M. M. (2018). Prefrontal cortex as a meta-reinforcement learning system. *Nature Neuroscience, 21*(6), 860–868. https://doi.org/10.1038/s41593-018-0147-8

Watkins, C. J. C. H., & Dayan, P. (1992). Q-learning. *Machine Learning, 8*, 279–292.

Werbos, P. (1982). Applications of advannces in nonlinear sensitivity analysis. In *Proceedings of the 10th IFIP Conference* (pp. 762–770). Springer-Verlag.

White, C. N., Ratcliff, R., & Starns, J. J. (2011). Diffusion models of the flanker task: Discrete versus gradual attentional selection. *Cognitive Psychology, 63*(4), 210–238.

White, C. N., Ratcliff, R., Vasey, M. W., & McKoon, G. (2010). Using diffusion models to understand clinical disorders. *Journal of Mathematical Psychology, 54*(1), 39–52. https://doi.org/10.1016/j.jmp.2010.01.004

Whittington, J. C. R., Muller, T. H., Mark, S., Chen, G., Barry, C., Burgess, N., & Behrens, T. E. J. (2020). The Tolman-Eichenbaum machine: Unifying space and relational memory through generalization in the hippocampal formation. *Cell, 183*(5), 1249–1263.

Widrow, G., & Hoff, M. E. (1960). Adaptive switching circuits. In *IRE Western Electric Show and Convention Record (pp. 96-104).* Lawrence Erlsbaum Associates.

Wiecki, T. V, Sofer, I., & Frank, M. J. (2013). HDDM: Hierarchical Bayesian estimation of the drift-diffusion model in Python. *Frontiers in Neuroinformatics, 7*(August), 1–10. https://doi.org/10.3389/fninf.2013.00014

Wilken, P., & Ma, W. J. (2004). A detection theory account of change detection. *Journal of Vision, 4*, 1120–1135. https://doi.org/10.1167/4.12.11

Williams, R. J. (1992). Simple statistical gradient-following algorithms for connectionist reinforcement learning. *Machine Learning, 8*, 229–256.

Wilson, R. C., & Collins, A. G. E. (2016). Ten simple rules for the computational modeling of behavioral data. *ELife*, 1–33.

Wilson, R. C., & Niv, Y. (2015). Is model fitting necessary for model-based fMRI? *PLoS Computational Biology*, 1–21. https://doi.org/10.1371/journal.pcbi.1004237

Wolpert, D. M. (2012). *The real reason for brains.*

Wolpert, D. M., Ghahramani, Z., & Jordan, L. M. (1995). An internal model for sensorimotor integration. *Science, 269*, 1880–1882.

Wong, K.-F., & Wang, X.-J. (2006). A recurrent network mechanism of time integration in perceptual decisions. *The Journal of Neuroscience, 26*(4), 1314–1328. https://doi.org/10.1523/JNEUROSCI.3733-05.2006

Wu, C. M., Schulz, E., Speekenbrink, M., Nelson, J. D., & Meder, B. (2018). Generalization guides human exploration in vast decision spaces. *Nature Human Behaviour, 2*, 915–924. https://doi.org/10.1038/s41562-018-0467-4

Wyart, V., de Gardelle, V., Scholl, J., & Summerfield, C. (2012). Rhythmic fluctuations in evidence accumulation during decision making in the human brain. *Neuron, 76*(4), 847–858. https://doi.org/10.1016/j.neuron.2012.09.015

Yu, A. J., & Dayan, P. (2005). Uncertainty, neuromodulation, and attention. *Neuron, 46*(4), 681–692. https://doi.org/10.1016/j.neuron.2005.04.026

Zhang, W., & Luck, S. J. (2008). Discrete fixed-resolution representations in visual working memory. *Nature, 453*(7192), 233–235. https://doi.org/10.1038/nature06860

Zylberberg, A., Fetsch, C. R., & Shadlen, M. N. (2016). The influence of evidence volatility on choice, reaction time and confidence in a perceptual decision. *ELife, 5*(October), 1–31. https://doi.org/10.7554/eLife.17688

Index

Activation function
  linear, 53
  logistic, 54
  sigmoid (*see* Activation function, logistic)
  threshold, 54
Actor-critic algorithm, 131
Anderson, James, 41–44
Anderson, John, 184, 186
Approximate Bayesian computation, 95–96
Attention, 130–131, 186–187
Attractor, 29
Auto-encoder, 161–162

Backpropagation
  basic algorithm, 72–74
  in Reinforcement Learning, 129
  through time, 79
Bahrami, Bahador, 193
Bayesian statistics, 173–176
Bellman equations, 139
Binding problem, 87
Blocking, 62
Bogacz, Rafal, 188–189
Boltzmann Machine, 162–166
  Restricted, 166–167
Bootstrapping, 99
Botvinick, Matthew, 3
Burgess, Neil, 50–51
Busemeyer, Jerome, 111

Catastrophic interference, 75–76
Categorization, 82, 184–186
Christiansen, Morten, 201
Cohen, Jonathan, 3

Coherent coactivation, 84
Collins, Anne, 116
Competitive learning, 156–158
Compositionality, 86–87
Computational psychiatry, 103, 105
Connection. *See* Weight
Convex set, 71
Cooperation, 195–198
Credit assignment, 73, 130, 142
Crockett, Molly, 192–193
Cross-entropy function, 57
Cross-validation, 119–120
Cultural evolution, 199
Curse of dimensionality, 90, 177, 181, 186

D'Ardenne, Kimberlee, 146
Dayan, Peter, 169–170
Diffusion model
  applications of, 35–36
  mechanism of, 33–35
  parameter estimation of, 101–102
Distributed information, 63, 80, 86
Distributions (Prior and Posterior), 174
Dobzhansky, Theodosius, 6
Dopamine, 130, 145–146
Drift rate, 34
Dynamic programming, 138–140

Effect
  distance, 23–24
  frequency, 49
  primacy, 50
  recency, 50
  size, 24

Egner, Tobias, 170–171
Elman, Jeffrey, 79–80
Energy function, 21, 28, 38
Equilibrium, 165, 166
Expected utility theory, 9
Exploration-exploitation dilemma, 126, 143–144

Feedback alignment, 88
Franconeri, Steven, 161
Friston, Karl, 164, 188–189

Game theory, 193–195
Gating, 149
Generalization, 86–87
Generative model, 169
Gergely, György, 188
Goal function, 7–12
Gradient descent
  basic principle of, 10–12
  gradient ascent versus, 12
  stochastic, 55, 83
Gridworld, 136–138

Hare, Mary, 199–200
Harmony theory, 30
Hauser, Tobias, 121
Helmholtz, Hermann, 169, 182
Helmholtz machine, 169–170
Hidden unit, 69–72
Hinton, Geoffrey, 167–169
Hippocampus, 32–33
Holroyd, Clay, 115, 147
Homeostasis, 124
Homo economicus, 8
Hopfield model, 27–30, 44–48
Hutcherson, Cendri, 192
Hyperparameters, 101

Ideal Observer model, 184
Information Criterion
  Akaike, 117
  Bayesian, 117
Interactive Activation Model, 25–27

Jara-Ettinger, Julian, 187–188
Joint modeling, 102

Kahneman, Daniel, 9–10
Kandel, Erik, 40–41

Keramati, Mehdi, 124
Kirby, Simon, 200
Körding, Konrad, 183

Latent variable, 185
Learning rule
  competitive, 156–158
  deep, 70
  delta, 53–55
  Hebbian, 37–40
  iterated, 198–199
  Kohonen, 158–160
  local availability of information in, 40, 88
  Rescorla-Wagner, 62, 140–141
  supervised, 37
  temporal differences, 141–142
  unsupervised, 153
  unsupervised Hebbian, 153–156
Likelihood
  function, 92–93, 174
  method of maximum, 56–57, 93–96, 164
Linear independence, 58
Linear principle, 19
Linear separability, 58
Linsker, Gary, 155–156
Local minimum, 74–75

Machine learning, 17–18
Markov decision process, 134–135
Marr, David, 15
McClelland, James, 25–27
McRae, Ken, 48
Meyniel, Florent, 183
Minimization
  energy, 21–22, 38
  loss, 180
  mean squared error, 55–56
Minsky, Marvin, 66–67
Model-based EEG, 116
Model-based fMRI, 114–115
Model comparison
  via Akaike Information Criterion, 117–119
  via Bayesian Information Criterion, 117–118
  via cross-validation, 119–120
Model evaluation, 107
Mulder, Martijn, 36

N-armed bandit, 96–97, 126–127
Negativity (Error- and Feedback-related), 147

Index

245

Neural network
  convolutional, 77–79
  multilayer, 69–72
  radial basis function, 81–82
  recurrent, 79–81
  two-layer, 53–54
Newton, Isaac, 1–2
Nowak, Martin, 195–198
Numerical cognition, 65, 158, 170

Oberauer, Klaus, 120
Omnibus test, 108–109
On-policy and off-policy algorithms, 142
Oppenheim, Gary, 64–65
Optimistic initial values, 144
Optimization principle, 7–10
Orthogonality and -normality, 42

Palminteri, Stefano, 111–112
Parameter
  estimation by maximum likelihood, 92–96
  identifiability, 96–99
  inverse temperature, 97
  learning rate, 39
  recovery, 98
  scaling, 11
  space exploration, 89–91
Past tense formation, 53–54
Pattern completion, 32
Pattern separation, 32
Pet detector, 17–24
Policy
  basic principle of, 135
  gradient, 130–131
  greedy, 143–144
  iteration, 143
  updating, 143
Prediction error, 62, 72, 76, 79, 115, 116, 126
Predictive coding, 190
Principal component, 154–156
Prisoner's dilemma game, 194–196
Python, x

Q-learning, 142
Quasi-regularity, 64

Radial basis function network, 81–82
Ratcliff, Roger, 33
Reali, Florencia, 200–201

Rectified Linear Unit (RELU), 77–79
Regularization (of rules), 63, 200
Regularization (of weights), 74
REINFORCE algorithm, 128–129
Reinforcement Learning
  Model-based, 138–139
  Model-free, 141
Replay, 75–76, 150–151
Representational dissimilarity matrix, 115
Resampling method, 113–114
Retrieval-induced forgetting, 64–65
Roelfsema, Pieter, 130
Rogers, Timothy, 84–85
Rumelhart, David, 63

Semantic cognition, 83–85
Short-term memory, 50–51, 99–101, 161
Social decision making, 192–193
Softmax choice rule, 97, 144
Stroop task, 3, 67
Subsampling layer, 79
Sutton, Richard, 133

TensorFlow, 13
Testolin, Alberto, 170

Universal grammar, 201
Upper confidence bound method, 145

Value
  concept of, 137
  estimation, 138–142
van den Berg, Ronald, 120
Verguts, Tom, 65, 158

Wagenmakers, Eric-Jan, 118–119
Weight, 19
Weight sharing, 77
White, Corey, 36
Williams, Ronald, 127–129
Word naming, 64–65
Working memory. *See* Short-term memory

Yu, Angela, 186–187